Lecture Notes in Networks and Systems

Volume 53

Series editor

Janusz Kacprzyk, Polish Academy of Sciences, Warsaw, Poland
e-mail: kacprzyk@ibspan.waw.pl

The series "Lecture Notes in Networks and Systems" publishes the latest developments in Networks and Systems—quickly, informally and with high quality. Original research reported in proceedings and post-proceedings represents the core of LNNS.

Volumes published in LNNS embrace all aspects and subfields of, as well as new challenges in, Networks and Systems.

The series contains proceedings and edited volumes in systems and networks, spanning the areas of Cyber-Physical Systems, Autonomous Systems, Sensor Networks, Control Systems, Energy Systems, Automotive Systems, Biological Systems, Vehicular Networking and Connected Vehicles, Aerospace Systems, Automation, Manufacturing, Smart Grids, Nonlinear Systems, Power Systems, Robotics, Social Systems, Economic Systems and other. Of particular value to both the contributors and the readership are the short publication timeframe and the world-wide distribution and exposure which enable both a wide and rapid dissemination of research output.

The series covers the theory, applications, and perspectives on the state of the art and future developments relevant to systems and networks, decision making, control, complex processes and related areas, as embedded in the fields of interdisciplinary and applied sciences, engineering, computer science, physics, economics, social, and life sciences, as well as the paradigms and methodologies behind them.

Advisory Board

Fernando Gomide, Department of Computer Engineering and Automation—DCA, School of Electrical and Computer Engineering—FEEC, University of Campinas—UNICAMP, São Paulo, Brazil
e-mail: gomide@dca.fee.unicamp.br

Okyay Kaynak, Department of Electrical and Electronic Engineering, Bogazici University, Istanbul, Turkey
e-mail: okyay.kaynak@boun.edu.tr

Derong Liu, Department of Electrical and Computer Engineering, University of Illinois at Chicago, Chicago, USA and Institute of Automation, Chinese Academy of Sciences, Beijing, China
e-mail: derong@uic.edu

Witold Pedrycz, Department of Electrical and Computer Engineering, University of Alberta, Alberta, Canada and Systems Research Institute, Polish Academy of Sciences, Warsaw, Poland
e-mail: wpedrycz@ualberta.ca

Marios M. Polycarpou, KIOS Research Center for Intelligent Systems and Networks, Department of Electrical and Computer Engineering, University of Cyprus, Nicosia, Cyprus
e-mail: mpolycar@ucy.ac.cy

Imre J. Rudas, Óbuda University, Budapest Hungary
e-mail: rudas@uni-obuda.hu

Jun Wang, Department of Computer Science, City University of Hong Kong
Kowloon, Hong Kong
e-mail: jwang.cs@cityu.edu.hk

More information about this series at http://www.springer.com/series/15179

Giedrius Laukaitis
Editor

Recent Advances in Technology Research and Education

Proceedings of the 17th International
Conference on Global Research
and Education, Inter-Academia – 2018

 Springer

Editor
Giedrius Laukaitis
Department of Physics,
   Faculty of Mathematics
   and Natural Sciences
Kaunas University of Technology
Kaunas, Lithuania

ISSN 2367-3370   ISSN 2367-3389   (electronic)
Lecture Notes in Networks and Systems
ISBN 978-3-319-99833-6   ISBN 978-3-319-99834-3   (eBook)
https://doi.org/10.1007/978-3-319-99834-3

Library of Congress Control Number: 2018952236

This Springer imprint is published by the registered company Springer Nature Switzerland AG
The registered company address is: Gewerbestrasse 11, 6330 Cham, Switzerland

Preface

It is commonly admitted at present that the border between the basic and applied researches in sciences becomes more and more diffuse and time lag between the fundamental discoveries and their implementation in technology and social life diminishes continuously with an impressive rate. This is associated with an unprecedented interdisciplinary approach and innovative tools for the progress of both basic knowledge and emerging technologies. Among the most spectacular breakthroughs of the technologies developed during the last decade are grounded on the interconnected fields of natural and life sciences, mathematics, computer science, and engineering.

This volume of proceedings is the outcome of the 17th International Conference on Global Research and Education, Inter-Academia 2018 hosted by Kaunas University of Technology, Lithuania, from September 24 to 27, 2018, and organized by the physics department of Faculty of Mathematics and Natural Sciences of Kaunas University of Technology (https://interacademia.ktu.edu). It is the fourth volume in the series, following the editions from 2015 to 2017. The International Advisory Conference Committee was run by the members of the International Society of Inter-Academia community.

The *International Conference on Global Research and Education, Inter-Academia* is an annual event organized every autumn by the Inter-Academia community since 2002. The main goal of Inter-Academia 2018 is to provide a well-articulated international forum to review, stimulate, and understand the recent trends in both fundamental and applied researches, along with the associated educational programs and training for students. Inter-Academia community (I-AC) —a body comprising 14 universities and research institutes from Japan and Central/Eastern European countries that agreed, in 2002, to coordinate their research and education programs so as to better address today's challenges. This conference promotes the Inter-Academia philosophy among professors, researchers, and students from the partner institutions—Shizuoka University, East-Central and Eastern European universities. Besides the interest in recent research results, the conference aims to strengthen the cooperation between the partners of

Inter-Academia community toward new theoretical and practical advances in research.

The Inter-Academia 2018 international conference continues a series of 16 uninterrupted similar events, initiated in 2002 and successively hosted by the partner universities. This event is an exceptional platform that brings together a wide spectrum of thematic areas in the fields of material science and technology, nanotechnology and nanometrology, biotechnology and environmental engineering, plasma physics, photonics, manufacturing technology, signal and image processing, electric and electronic engineering, measurement, identification, and control, intelligent and soft computing techniques, modeling and diagnostics, robotics, precision engineering, Internet-based education, distance learning, and multimedia and e-learning.

Thirty-eight selected papers were included in the Springer proceedings after a thorough two-stage review and editing process. All the manuscripts submitted to the Inter-Academia 2018 were peer-reviewed by at least two independent reviewers, who were provided with a detailed review form. The comments from the reviewers were communicated to the authors, who incorporated the suggestions in their revised manuscripts. The recommendations from reviewers were taken into consideration while selecting a manuscript for inclusion in the proceedings. Herein, we would like to express our gratitude to the authors, and especially to reviewers, who greatly contributed to ensuring the proper scientific level of the articles in this book.

On behalf of the organizing committee, we would like to express our thanks to the conference's International Scientific Committee and the Advisory Committee, as well as all those who have contributed to this conference, for their support and advice.

Giedrius Laukaitis

Contents

Material Science and Technology, Smart Materials, Nanotechnology

Structure and Mechanical Properties
of Gradient Metal-Carbon Coatings

Alexandr V. Rogachev[1,2]($\boxtimes$), Ekaterina A. Kulesh[2],
Dmitry G. Piliptsou[1,2], Alexandr S. Rudenkov[1,2], and Jiang X. Hong[1]

[1] International Chinese-Belarusian Scientific Laboratory on Vacuum-Plasma
Technology, Nanjing University of Science and Technology, Xiao Ling Wei 200,
Nanjing 210094, China
[2] Francisk Skorina Gomel State University, Sovetskaya 102,
246019 Gomel, Belarus
rogachevav@mail.ru

Abstract. Gradient coatings were obtained as a result of deposition from the combined titanium plasma fluxes of a DC arc discharge and the flow of carbon ions formed by sputtering a graphite target with an impulse arc discharge. Structure and surface morphology of the coatings were analyzed by Raman spectroscopy and AFM. Mechanical properties were studied by friction (a "sphere-plane" method) and microhardness test (AFFRI DM8 test machine with a Knoop style diamond tip). The dependence between the carbon concentration in the depth of the studied films and their microhardness has been found out. Raman spectroscopy showed that films with high contents of the sp^3 phase are formed, the presence of titanium atoms leads to a decrease the size of the carbon sp^2 cluster, nitrogen conducts to an increase in the degree of disordering of s^2 carbon clusters. It is established that the parameters of friction and wear depend on the structure of the gradient films.

Keywords: Carbon coatings · Gradient coatings · Raman spectroscopy
Microhardness · Friction and wear

1 Introduction

Amorphous carbon (a-C) coatings are widely used as wear-resistant protective coatings for friction units and tools [1]. It is known [2] that the properties of coatings, such as microhardness, friction, internal stresses are determined by the structure, namely the ratio of sp2/sp3 components in the coating. Improving the mechanical properties of carbon coatings is achieved both by optimizing the ion-plasma coating application and by changing the structure of the sublayer and the main amorphous carbon layer of the multilayer coating, including the co-building of the layers by periodically changing the contents of various elements on the thickness of the layer. Gradient coatings containing layers of amorphous carbon are used to reduce the difference in the coefficient of thermal expansion between different layers of coating and substrate and to increase the tribological properties of materials.

© Springer Nature Switzerland AG 2019
G. Laukaitis (Ed.): INTER-ACADEMIA 2018, LNNS 53, pp. 3–10, 2019.
https://doi.org/10.1007/978-3-319-99834-3_1

The idea of creating gradient layers based on carbon coating was first put forward with the goal of obtaining materials that can withstand heavy thermomechanical loads arising from the operation of aviation equipment and diesel engines. Developed coatings should combine hardness, wear-resistance of "diamond-like" coatings with plasticity of metals. A gradual change in the microstructure without sharp interfaces, a smooth change in the microhardness, and a convergence of the elastic modulus of the "diamond" and metal layers would lead to an increase in the strength of the coating and its longevity.

2   Experimental

For the formation of gradient coatings based on carbon and titanium, it is proposed to use a vacuum device with a pulsed arc source of a carbon plasma, an DC arc source of a metal plasma, and an ion source. Figure 1 show a schematic diagram of the experiment.

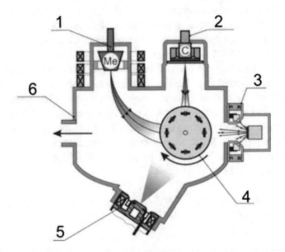

Fig. 1. Scheme of a vacuum device for the formation of gradient and coatings: 1-DC arc of Ti plasma, 2-pulse arc carbon plasma evaporator, 3-sputtering ion source, 4-rotating table with samples, 5-ion source for assistance, 6-vacuum chamber

The use of composite plasma streams of carbon and metal has several advantages and allows:

1. control the elemental composition of coatings,
2. control the structure of the coating, namely, the ratio of carbon atoms to the different hybridization of bonds,
3. to carry out a chemical interaction between the atoms of metal, nitrogen and carbon.

The formation of carbide or carbonitride compounds occurs both as a result of the interaction in the plasma flow and on the substrate. Due to the change in the rates of metal deposition (regulated by changing the discharge current) and the carbon plasma (regulated by changing the pulse repetition frequency), it is possible to control the ratio of the metallic and carbon components in the coating. By changing the ratio of the number of carbon atoms and the metal deposited on the substrate, it is possible to change the concentration of the elements in the coating. Typically, the deposition of the gradient coating begins with deposition of the metal layer, which allows to provide the necessary values of the adhesive bond with the substrate material. The next stage is the introduction of carbon atoms into the metallic plasma flow and the subsequent growth of the concentration of carbon atoms up to 100% at. in the upper layer. A regulated change in the carbon concentration will allow the formation of Me-C bonds, which leads to a decrease in internal stresses in the coating and allows deposition of coatings up to 1 μm in thickness. Figure 2 shows a scheme of gradient coatings.

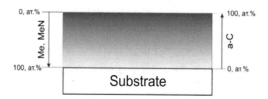

Fig. 2. Scheme of gradient coatings

In addition, in order to change the structure of the coatings and to realize the plasma-chemical reactions in the stages of growth and deposition of the layers, the cathode material was sputtered in a molecular nitrogen atmosphere at a nitrogen partial pressure of 10–2 Pa. Coatings were obtained in the following deposition modes:

1. Ti/a-C (5 … 20 Hz): Ti- on a titanium sublayer with a 120 nm thick, nanocomposite carbon-titanium layers were deposited, with an increase in the repetition frequency of the discharge pulses of a carbon source from 5 to 20 Hz in steps of 5 Hz and arc-arc current with cathode from titanium 80 A, which allowed to realize a change in the C/Ti ratio from 0.1 to 1.4.
2. Ti/a-C (5 … 20 Hz): Ti/a-C - difference from the sample Ti/a-C(5 … 20 Hz):Ti in the presence of the upper layer a-C.
3. Ti/a-C(5 … 20 Hz): Ti:N_2/a-C:N_2Ti/a-C deposition (5 … 20 Hz):Ti/a-C was carried out in a molecular nitrogen atmosphere with base pressure 10^{-2} Pa.

The notation (5 … 20 Hz) shows that the repetition rate of the pulses of the carbon plasma source varied from a minimum value of 5 Hz to a maximum value of 20 Hz. By changing the frequency, the ratio of carbon and titanium concentrations in the coating changes.

The microstructure of the coatings was studied by Raman spectroscopy using a Senterra microscope (Bruker). The spectra were excited by the laser radiation with wavelength 532 nm and of 20 mW power. Raman spectroscopy is used as an effective method for studying the microstructure of amorphous carbon coatings, and allows determining the presence of certain types of bonds between carbon atoms and the dimensions of carbon clusters [3].

The microhardness of the coatings was determined by the Knoop method in 15 different places on the surface of the coatings, the average value was determined from the results obtained. The load and duration of the test were 245 mN and 10 s, respectively. The hardness of the silicon substrate measured in this mode was 9.56 ± 0.32 GPa.

Surface morphology and roughness (Ra) were studied by the Solver 47 (NT-MDT) atomic force microscope (AFM) in tapping mode with a scanning scope of 10×10 μm and a scan rate of 0.1 μm/s.

Tribological tests of gradient coatings were carried out using a "sphere-plane" method with 478 MPa contact pressure at 23°C and 70% relative humidity and as a counter body was used a ball with a diameter of 5 mm made of steel ШХ 15.

3   Result and Discussion

Figure 3 shows the Raman spectra for coatings with different contents of the carbon component over the thickness of the layer. For all coatings, there is a wide asymmetric peak in the region of 1000–1800 cm^{-1}.

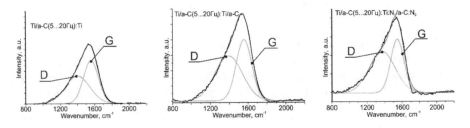

Fig. 3. Raman spectra of gradient coatings

Raman spectra of gradient carbon coatings (Fig. 3) are characterized by the presence of two components, namely the D peak with the center at 1368 cm^{-1} and the G peak with the center at 1550 cm^{-1}. The ratio of the integrated intensity ID/IG of these peaks is given in Table 1 and determines the relative content of carbon atoms with sp2 and sp3 hybridization of bonds [1, 3]. As is known [4], the sp2 state is characteristic for carbon atoms forming graphite, the presence of which in the coating makes it possible to reduce friction in the contact zone, and the sp3 state is characteristic for carbon atoms forming the crystal lattice of diamond, which determines the strength and hardness of amorphous carbon coatings. The presence of impurities or alloying elements leads to a change in the sp2/sp3 ratio of the carbon atoms in the coating.

G peak in the Raman spectra is formed due to the vibrations of pairs of carbon atoms in the sp^2 state in rings and linear chains. The peak D is formed due to the so-called "breath" modes of carbon atoms with sp^2 hybridization bonds in aromatic rings [1, 4]. The position and width of the peak G determine the change in the size and degree of ordering of the carbon Csp^2 clusters.

Table 1. Fitting results of Raman spectra of gradient coatings

Samples	I_D/I_G ratio	G peak shift, cm^{-1}	G peak weight, cm^{-1}
Ti/α-C(5...20 Hz):Ti	1.05	1556.8	173.2
Ti/α-C(5...20 Hz):Ti/α-C	1.29	1556.8	165.7
Ti/α-C(5...20 Hz):Ti:N₂/α-C:N₂	1.49	1545.8	153.4

Depending on the deposition parameters, the presence of carbon or nitrogen atoms in the carbon matrix changes the shape and width of the Raman spectra (Fig. 3). It has been established (Table 1) that when molecular nitrogen dopes the gradient titanium-carbon coating, the ID/IG ratio increases to 1.49, which determines the increase in the number of carbon atoms in the state with sp2 hybridization of bonds, and also by the formation of CN and C=N, which leads to a decrease in the size of the Csp2 clusters (a decrease in the width of the G peak compared to Ti/α-C(5...20 Hz):Ti/α-C coating), as well as a decrease in the degree of disorder of the Csp2 clusters, which is confirmed by shifting the position of G peak to the region of low wavenumbers.

Taking into account the thickness of the upper α-C: N_2, as well as the position and width of the peak G, we can draw the following conclusions:

1. α-C:N_2 layer consists of a composition of nanocrystalline graphite (NC graphite) and amorphous (diamond-like) carbon (α-C).
2. The introduction of nitrogen atoms can cause two independent processes-an increase in the sp^3 fraction and regulation of the sp^2 phase.

As shown in [1], when excitation of Raman spectra with visible-range radiation, an increase in the number of Csp^2 clusters results in a shift in the position of peak G to the region of high wave numbers, but in our case, the peak position G shifted to the region of lower wavenumbers up to $\sim$ 1545.8 cm^{-1} with a simultaneous decrease in the width G of the peak. This behavior of the G peak in Ti/α-C(5...20 Hz):Ti:N_2/α-C:N_2 coating indicates that not only the N atoms cause clustering of sp^2 carbon atoms, but also the formation of chemical bonds of the type Ti–N, Ti-C, Ti–C–N [5].

As shown in Table 1, for Ti/α-C(5 ... 20 Hz):Ti coating, the I_D/I_G ratio is reduced to a value of 1.05 and the G peak width. It is known [1, 3], that the concentration of carbon atoms in the sp^3 hybridization state is inversely proportional to the I_D/I_G ratio, which, according to Table 1, shows the effect of Ti atoms in Ti/α-C(5 ... 20 Hz):Ti coatings on growth of the sp^3 phase [6]. An increase in the con-centration of sp^3 bonds in Ti/α-C(5 ... 20 Hz):Ti coating leads to an increase in the degree of disordering of Csp^2 clusters in the carbon matrix, which determines the broadening of the G peak.

The fitting result of the Raman spectrum of the Ti/α-C(5 ... 20 Hz):Ti/α-C coatings show values typical for a monolayer of α-C, which suggests a weak effect of gradient α-C(5... 20 Hz):Ti layers on the processes of structure formation in the upper α-C layer.

Figure 4 shows the topography of the surface of the coatings (a), the image of the phase contrast (b) and the particle size distribution (c). The images allow us to fully characterize the morphology of the surface of coatings and to establish the features of growth as a function of the structure of coatings [7].

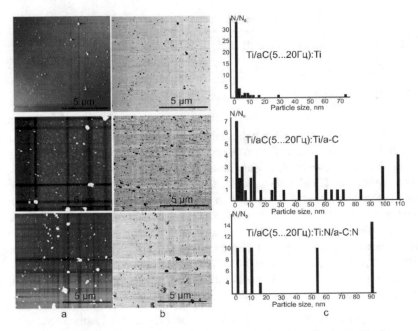

Fig. 4. AFM images of the gradient coatings based on carbon, titanium and nitrogen

According to Fig. 4a, it can be concluded that the obtained coatings are dense without pores and peelings [8]. The surface of the coatings, regardless of the architecture of the coating, looks smooth, but on the surface of Ti/a-C(5...20 Гц):Ti and Ti/a-C(5...20Гц):Ti:N$_2$/a-C:N$_2$ coatings, single grains of large size are present, which is caused by deposition regimes and the presence of a titanium drop phase in the ion flow.

The images of phase contrast (Fig. 4b) show the relative distribution of mechanical properties over the surface, namely hardness (more dark areas correspond to areas with lower hardness) and shows that for all gradient coatings, droplet formations on the surface have a hardness lower than the main matrix, and as was established in previous works are particles of graphite or titanium [9].

The distribution of grains on the surface is shown in Fig. 4c. The particle size distribution showed that the most probable particle size lies in the region up to 10 nm, but there are also single droplets up to 90 nm in size that arise when a plasma stream is generated, a titanium particle or a fragment of a graphite cathode.

The results of treatment of AFM images (Table 2) such as Ra (average roughness) and d (average diameter of grain) show the features of surface formation depending on the presence of nitrogen and/or titanium atoms in the coating [10].

It is shown that the Ti/a-C(5...20 Hz):Ti/a-C coating has a minimum roughness, which agrees with the data obtained earlier on the morphology of a-C coatings obtained at high frequencies of the discharge pulses (high density of the plasma flow). The presence of titanium atoms in the coating leads to an increase in roughness, which is determined by the features of the generation of the titanium ion flux and the presence of a drop phase in it. Due to the presence of a dense stream of titanium ions and the etching of the growing carbon coating by the high-energy component of the plasma flow of titanium ions, the average grain diameter of the Ti/a-C(5...20 Hz):Ti coating decreases, forming a denser structure consisting of carbon in the sp3 state and titanium carbide.

Table 2. AFM parameters of the morphology of gradient coatings

Samples	Ra, nm	d, nm
Ti/a-C(5...20 Hz):Ti	6.7	2.6
Ti/a-C(5...20 Hz):Ti/a-C	2.3	12.8
Ti/a-C(5...20 Hz):Ti:N_2/a-C:N_2	3.5	3.4

The results of determining the mechanical properties, namely microhardness and friction, are given in Table 3.

Table 3. Mechanical properties of gradient coatings

Samples	Hv, GPa	Friction coefficient μ	Counterbody wear rate, $j \times 10^{-11}$, m^3/(N·m)
Ti/a-C(5...20 Hz):Ti	10.7	0.24	3.3
Ti/a-C(5...20 Hz):Ti/a-C	11.6	0.22	2.9
Ti/a-C(5...20 Hz):Ti:N_2/a-C:N_2	12.8	0.18	2.4

As can be seen from the data given in Table 3, the microhardness depends on the coating architecture, the presence of nitrogen and titanium atoms in the coating. The hardest coating is the coating containing nitrogen in its composition: it is determined by the presence of a phase of titanium nitride, and also by a decrease in the size of the sp2 cluster of carbon.

As can be seen from Table 3 the friction coefficient and counterbody wear rate of the Ti/a-C(5...20 Hz):Ti:N_2/a-C:N_2 is considerably lower compared with that of the non-doped with nitrogen coatings. For the Ti/a-C(5...20 Hz):Ti coatings an increase of the friction coefficient up to 0.24 is observed, that can be attributed to the increase in Ra (see Table 2) and the decrease in microhardness (see Table 3).

On the one hand, the presence of nitrogen atoms in the structure of Ti/a-C(5...20 Hz):Ti:N2/a-C:N2 coating decreases the content of sp^3 carbon bonds and stabilizes the Csp^2 clusters structure, increasing the microhardness of the coating by forming TiN phase in coating.. On the other side, a lower coefficient of friction of the Ti/a-C(5...20 Hz):Ti:N2/a-C:N2 can be additionally explained by the formation of a lubricating graphite surface layer [12].

4 Conclusions

The dependence between the structure of gradient coatings and their microstructure, surface morphology and mechanical properties has been found out. Raman spectroscopy showed that films with high contents of the sp^3 phase are formed, the presence of titanium atoms leads to a decrease the size of the carbon sp^2 cluster, nitrogen conducts to an increase in the degree of disordering of s^2 carbon clusters. The N-doping of uplayer of the gradient coatings leads to a decrease in the friction coefficient and substantial changes in both the phase compositions and surface morphology, which, in turn, reduce the counterbody wear rate and noticeably increase microhardness of these coatings, due to the formation nitride layers in the volume of coating.

Acknowledgments. This work was supported by The Belarusian Republican Foundation for Research (project No. T18-KI08, for 2018-2019) and Intergovernmental Cooperation Projects in the National Key Research and Development Plan of the Ministry of Science and Technology of PRC (projects No. 2016YFE0111800, for 2016–2019).

References

1. Robertson, J.: Diamond-like amorphous carbon. Mater. Sci. Eng. **37**, 129 (2002)
2. Ren, Y., Erdmann, I., Küzün, B., Deuerler, F., Buck, V.: Effect of deposition parameters on wear particle size distribution of DLC coatings. Diam. Relat. Mater. **23**, 184–188 (2012)
3. Ferrari, A.C., Robertson, J.: Interpretation of Raman spectra of disordered and amorphous carbon. Phys. Rev. B **61**, 14095–14107 (2000)
4. Piscanec, S., Lazzeri, M., Mauri, F., Ferrari, A.C., Robertson, J.: Kohn anomalies and electron-phonon interactions in graphite. Phys. Rev. Lett. **93**, 185503 (2004)
5. Donnet, C., Erdemir, A.: Tribology of diamond-like carbon films. In: Ferrari, A.C. (ed.) Non-destructive Characterisation of Carbon Films, pp. 25–82. Springer, New York (2008)
6. Cui, W.G., Lai, Q.B., Zhang, L., Wang, F.M.: Quantitative measurements of sp3 content in DLC films with Raman spectroscopy. Surf. Coat. Technol. **205**, 1995–1999 (2010)
7. Kocourek, T., Růžek, M., Landa, M., Jelínek, M., Mikšovský, J., Kopeček, J.: Evaluation of elastic properties of DLC layers using resonant ultrasound spectroscopy and AFM nanoindentation. Surf. Coat. Technol. **205**, S67–S70 (2011)
8. Yi, M., Piliptsou, D.G., Rudenkov, A.S., Rogachev, A.V., Jiang, X., Dongping, S., Chaus, A.S., Balmakou, A.: Structure, mechanical and tribological properties of Ti-doped amorphous carbon fi lms simultaneously deposited by magnetron sputtering and pulse cathodic arc. Diam. Relat. Mater. **77**, 1–9 (2017)
9. Yi, M., Jiang, X., Piliptsou, D.G., Zhuang, Y.: Chromium-modified a-C films with advanced structural, mechanical and corrosive-resistant characteristics. Appl. Surf. Sci. **379**, 424–432 (2016)
10. Liu, Z.H., Lemoine, P., Zhao, J.F., Zhou, D.M., Mailley, S., McAdams, E.T., Maguire, P., McLaughlin, J.: Characterisation of ultra-thin DLC coatings by SEM/EDX, AFM and electrochemical techniques. Diam. Relat. Mater. **7**, 1059–1065 (1998)
11. Chaus, A.S., Jiang, X.H., Pokorný, P., Piliptsou, D.G., Rogachev, A.V.: Improving the mechanical property of amorphous carbon films by silicon doping. Diam. Relat. Mater. **82**, 137–142 (2018)

Recovery of Sintered Carbide Material in Electrochemical Machining Process

Sicong Wang[1], Akihiro Goto[2(✉)], Atsushi Nakata[2],
Kunio Hayakawa[1], and Katsuhiko Sakai[1]

[1] Shizuoka University, 3-5-1 Johoku, Naka-Ku, Hamamatsu, Japan
[2] Shizuoka Institute of Science and Technology,
Toyosawa, Fukuroi 2200-2, Japan
goto.akihiro@sist.ac.jp

Abstract. The authors have been developing the method to recover the material of sintered carbide in electrochemical machining (ECM). The ECM performance is affected by the change in the electrolyte ingredient when the sintered carbide material is dissolved into the electrolyte. Besides, tungsten and cobalt are costly rare metals. In this report, the feasibility of a process on recovery of tungsten and cobalt by changing soluble substances into insoluble ones was investigated, in which process tungsten as calcium tungstate and cobalt as cobalt hydrate. It was found that the tungsten and cobalt compounds could be recovered by the proposed processes.

Keywords: Electrochemical machining (ECM) · Sintered carbide
Recovery

1 Introduction

The use of hard sintered carbide material, whose main components are tungsten carbide (WC) and cobalt (Co), as a die material for metal forming processes is attracting attention as a method for increasing productivity. While it is expected that the use of sintered carbide for metal dies will increase in the future, sintered carbide is difficult to machine using normal cutting tools because of its hardness, and electrical discharge machining (EDM) has been the main method to process it. EDM has the drawbacks of low machining speed or the occurrence of cracks on the worked surface. On the other hand, there have been attempts to use cutting tools made of diamond or other special materials to directly machine sintered carbide. Yet, these methods also have problems such as slow machining speed and extremely high cost of the tools. Although much research is being performed in this field, the use of special tools to machine sintered carbide still has not achieved wide usage.

In this study, we deal with electrochemical machining (ECM). Although there have been previous studies on electrochemical machining of sintered carbide [1–12], it has not yet been adopted as a common technology. The conventional method of electrochemical machining is to immerse the workpiece and an electrode in an electrolyte, impart a negative polarity to the electrode, and allow electrical current to flow to remove material from the workpiece at a high rate, shaping it. In the case of sintered

© Springer Nature Switzerland AG 2019
G. Laukaitis (Ed.): INTER-ACADEMIA 2018, LNNS 53, pp. 11–19, 2019.
https://doi.org/10.1007/978-3-319-99834-3_2

carbide, however, it is necessary to also machine tungsten carbide, which is nonmetal material.

When a monopolar power source is used, NaOH or other substances must be added to the electrolyte to remove WO3, which is oxidation product of WC. Because of the addition of NaOH or other substances, the electrolyte becomes strong alkaline, and it brings the problem of safe operation. In order to solve this problem, a method to use neutral electrolyte and a bipolar power supply was proposed [1], Where WC is anodic oxidized to become WO3 when the electrode is negatively charged, then made to react with Na + to produce soluble Na2WO4 when the electrode is positively charged. It is a combination of a pure electrolytic phenomenon in which cobalt is removed, and a chemical reaction to remove the tungsten carbide.

In the ECM of sintered carbide, it is considered that the machining performance is affected by increasing the concentration of $WO_4{}^{2-}$ or Co^{2+} ions in the electrolyte as it is machined. Both W and Co, which are components of sintered carbide, are costly rare metals. Since there is concern about environmental problems and resource problems in recent years, it would be thought that technologies for recovering rare metals such as W and Co will be necessary in the future. In this paper, from the viewpoint of resource recycling and keeping the components of the electrolyte constant, we consider a process for the recovery of ECMed W and Co. Here, the method of recovering W and Co in the NaNO3 electrolyte and keeping the components of the electrolyte constant is investigated.

2   Collection Method of Tungsten

It is reported by Maeda et al. [1] that ECM of sintered carbide increases Na_2WO_4 in the electrolyte, and that Na^+ ions in the electrolyte are consumed and that ECM is eventually hindered due to the deterioration of electrolyte. Na_2WO_4 is a water-soluble substance, it is difficult to recover it in the ECM process. Here, in order to recover W and restore the components of the electrolyte, the following two methods are investigated.

The first method is to use the following chemical reactions used for $CaWO_4$ production [13].

$$Na_2WO_4 + Ca\,(NO_3)_2 \rightarrow CaWO_4 \downarrow + 2NaNO_3$$

When $Ca\,(NO_3)_2$ is added to Na_2WO_4 generated in ECM of sintered carbide, $CaWO_4$ is produced. $CaWO_4$ is insoluble in water, it is considered that it can be recovered because it precipitates.

The other method is to use the following reaction.

$$Na_2WO_4 + 2HNO_3 \rightarrow H_2WO_4 \downarrow + 2NaNO_3$$

When HNO_3 is added to Na_2WO_4 generated in ECM, H_2WO_4 is produced. It is also insoluble in water [14], and it is considered that H_2WO_4 can be recovered. H_2WO_4 is said to exist in the state of hydrate; $WO_3 \cdot H_2O$.

As a preliminary experiment, two liquids were prepared by dissolving 30 g of $NaNO_3$ and 10 g of $Na_2WO_4 \cdot 2H_2O$ in 200 ml of water, and a precipitation test was conducted. The liquid immediately after making mixing is colorless and transparent as shown in Fig. 1(a). 10 g of Ca $(NO_3)_2 \cdot 4H_2O$ was added to one of the two liquids. To the other liquid 5 ml of 60% HNO_3 was added. Figure 1b shows a photograph of a liquid with $Ca(NO_3)_2$, and Fig. 1(c) shows that with HNO_3. When Ca $(NO_3)_2$ was added to the liquid, it became cloudy in white, and when it was left for a while, a white substance precipitated. In the liquid to which HNO_3 was added, yellowish white substance precipitated. These liquids were filtered, and the precipitate was recovered. From the liquid to which Ca $(NO_3)_2$ was added, a white substance was recovered. From the liquid to which HNO_3 was added, yellowish white substance was recovered. After drying off yellowish white matter, it turned to yellow. The respective X-ray diffraction images are shown in Fig. 2. As shown in Fig. 2(a), it was confirmed that $CaWO_4$ was present in the white substance, and $WO_3 \cdot H_2O$ existed in the yellow substance as shown in Fig. 2(b).

(a) Mixed solution of $NaNO_3$ and Na_2WO_4 (b) Addition of Ca $(NO_3)_2$ (c) Addition of HNO_3

Fig. 1. The precipitation test of tungsten

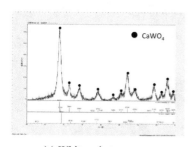 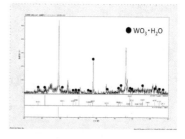

(a) White substance   (b) Yellowish white substance

Fig. 2. X-ray diffraction of precipitate including tungsten

3   Collection Method of Cobalt

In ECM of sintered carbide, As Co is eluted as Co^{2+} ions into the electrolyte, in the case of a sodium nitrate electrolyte, Co $(NO_3)_2$ is considered to be present in the electrolyte by the following reaction.

$$Co + 2NO^{3-} - 2e^- \rightarrow Co(NO_3)_2$$

Co $(NO_3)_2$ is a water-soluble substance, and it is difficult to recover it in the course of ECM process. Here, a method was tried to apply in order to recover Co that used the property that Co $(OH)_2$ is insoluble in water.

$$Co(NO_3)_2 + 2NaOH \rightarrow Co(OH)_2\downarrow + 2NaNO_3$$

When NaOH is added to Co $(NO_3)_2$ generated in ECM of sintered carbide, insoluble Co $(OH)_2$ is formed and precipitated, so that recovery is possible. A mixed liquid was prepared by dissolving 10 g of Co $(NO_3)_2\cdot6H_2O$ and 30 g of $NaNO_3$ in 200 ml of water. The state of the liquid was reddish brown transparent as shown in Fig. 3(a). Figure 3b shows a photograph when 5 g of NaOH was added to this solution. At the moment when NaOH was added, it turned blue. However, when leaving the liquid as it was, a pink substance precipitated in about 5 min, and the upper part of the liquid became colorless and transparent. The liquid was filtered, the pink substance was recovered, and X-ray diffraction was carried out. As shown in Fig. 4, it was confirmed that Co $(OH)_2$ existed in the pink material.

(a) Mixed solution of $NaNO_3$ and $Co(NO_3)_2$ (b) Addition of NaOH

Fig. 3. The precipitation test of Cobalt

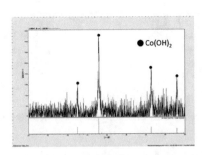

Fig. 4. X-ray diffraction of precipitate including Cobalt

4 Recovery Experiment from Liquid with Tungsten and Cobalt

The method of recovering W and Co from $NaNO_3 + Na_2WO_4$ mixed solution and $NaNO_3 + Co (NO_3)_2$ mixture solution was above mentioned. Next, assuming actual phenomenon of electrolyte in ECM of sintered carbide, the recovery experiment of W and Co simultaneously from $NaNO_3 + Na_2WO_4 + Co (NO_3)_2$ mixture liquid was conducted. First, 30 g of $NaNO_3$, 10 g of $Na_2WO_4 \cdot 2H_2O$ and 10 g of $Co (NO_3)_2 \cdot 6H_2O$ were dissolved in 200 ml of water, and we would have intended to add 5 g of NaOH to recover $Co (OH)_2$. However, before NaOH was added, when $NaNO_3$, Na_2WO_4 and Co $(NO_3)_2$ were mixed, purple substances precipitated as shown in Fig. 5(a). It is considered that $CoWO_4$ was generated by the following reaction [15].

$$Na_2WO_4 + Co(NO_3)_2 \rightarrow CoWO_4 \downarrow + 2NaNO_3$$

It was also found that this phenomenon occurs irrespective of the order to add the reagents. In spite the X-ray diffraction pattern of this violet substance was acquired, the substance could not be identified. It was confirmed by EDS (Energy dispersive X-ray spectrometry) that Co and W elements are contained. This liquid was filtered with a paper filter, and the obtained supernatant liquid was divided into two as shown in Fig. 5(b). 5 g of NaOH was added to the liquid on the left side and 10 g of Ca $(NO_3)_2 \cdot 4H_2O$ was added to the liquid on the right side. The result is shown in Fig. 5(c). A pink substance precipitated in the liquid on the left side, but the liquid on the right side did not change much, that is there was no $WO_4{}^{2-}$ ion in the supernatant. From this result, it was found that in the liquid with 10 g of $Na_2WO_4 \cdot 2H_2O$ and 10 g of Co $(NO_3)_2 \cdot 6H_2O$ amount of Co was excessive and all W in liquid was precipitated.

However, it is considered that more $WO_4{}^{2-}$ ions exists in the electrolyte than Co^{2+} ions in practical ECM process of sintered carbide. So, next, another tests were carried out with different amounts of Na_2WO_4 and Co $(NO_3)_2$. A solution was prepared by dissolving 30 g of $NaNO_3$, 15 g of $Na_2WO_4 \cdot 2H_2O$ and 5 g of Co $(NO_3)_2 \cdot 6H_2O$ in 200 ml of water. This time again as shown in Fig. 6(a), a purple substance was generated immediately after preparation. This liquid was filtered, and the supernatant liquid was divided into two as shown in Fig. 6(b). Unlike the previous experimental results, the supernatant liquid became colorless and transparent, not reddish brown. As in the previous experiment, 5 g of NaOH was added to the liquid on the left side and 10 g of $Ca(NO_3)_2 \cdot 4H_2O$ was added to the liquid on the right side. The result is shown in Fig. 5 (c). As the color of the liquid on the left was slightly changed, it can be considered that Co^{2+} ions were slightly present in the liquid. The liquid on the right became cloudy in white, and white substance precipitated over time. From this result, it was confirmed that W was present in the filtered liquid.

Originally, it was thought that when ECM of sintered carbide was carried out, the concentrations of $WO_4{}^{2-}$ and Co^{2+} would increase in the electrolyte and affect machining performance. However, from the results of experiments, it was found that when the concentrations of $WO_4{}^{2-}$ and Co^{2+} are increased, $CoWO_4$ is generated and precipitated and it can be removed from the electrolyte. Only the concentration of

WO_4^{2-} in the electrolytic solution becomes high, and it seems that only this substance should be recovered separately.

This experiment showed the possibility of recovering of W and Co in ECM of sintered carbide. However, this experiment was carried out with the reagents, not an actual ECM process. Next, this phenomenon was confirmed in real ECM process.

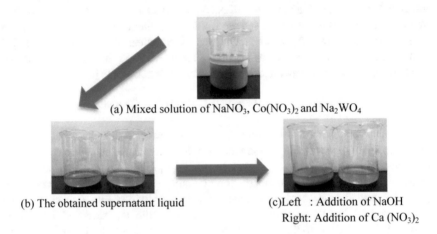

(a) Mixed solution of $NaNO_3$, $Co(NO_3)_2$ and Na_2WO_4

(b) The obtained supernatant liquid

(c)Left : Addition of NaOH
   Right: Addition of Ca $(NO_3)_2$

Fig. 5. Collection experiment of $NaNO_3$ + Na_2WO_4 + $Co(NO_3)_2$ mixture solution (1)

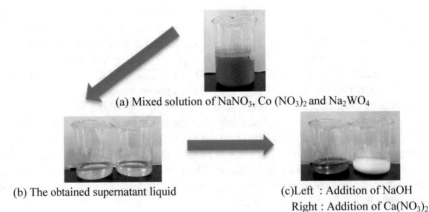

(a) Mixed solution of $NaNO_3$, Co $(NO_3)_2$ and Na_2WO_4

(b) The obtained supernatant liquid

(c)Left : Addition of NaOH
   Right : Addition of Ca$(NO_3)_2$

Fig. 6. Collection experiment of $NaNO_3$ + Na_2WO_4 + Co $(NO_3)_2$ mixture solution (2)

5   The Machining Experiment of Sintered Carbide by ECM

To verify the above mentioned result with reagents, actual ECM test of sintered carbide was we carried out. Figure 7 shows a schematic diagram of the experiment setup and the waveform of gap voltage and machining current. The electrode was a □ 10 mm 10 mm sized graphite that had a $\varphi 3$ mm hole thorough which electrolyte flowed. The workpiece was a sintered carbide with a worked surface of □ 11 mm × 11 mm. In this experiment, $T_1 = T_2 = 10$ ms. Currents I_1 and I_2 were both approximately 10 A. The electrolyte was about 13% $NaNO_3$ aqueous solution (1800 g $NaNO_3$ and 12 l water). To machine the workpiece, the initial distance between the workpiece and electrode was set at 0.1 mm or slightly higher, and the electrode was moved closer manually to the workpiece whenever the current flow had become reduced by 10–20%. Approximately 80 g of sintered carbide was machined. After machining, the electrolyte concentration became about 14.2%, which was caused by the evaporation of water.

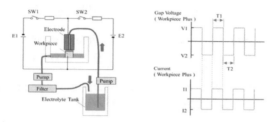

Fig. 7.   Outline of experiment setup

In the initial stage of machining, the filtration device was not added in the electrolyte circulation. When the sintered carbide was machined about 20 g, a phenomenon in which discharge occurred had been observed. We thought that the cause of discharge was the debris of ECM. Then, by using a filter in the electrolyte circulation, the machining was carried out without discharge. The electrolyte used in this experiment was filtered and the precipitate was dried. Though the precipitate of the electrolyte was tried to analyze with XRD it could not be identified probably because the crystal was small. But with EDS, Co and W was detected. It can be inferred that the same reaction as the experiment using the reagent occurred.

(a) after machining   (b) after filtering   (c) after adding chemicals

Fig. 8.   Electrolyte recovery experiment

After the filtration of the supernatant was divided into two portions as shown in Fig. 8(b). 5 g NaOH was added to the left solution and 10 g Ca $(NO_3)_2 \cdot 4H_2O$ was added to the right one. The appearance after adding reagent is shown in Fig. 8(c). Both precipitates were also filtered, dried and analyzed. From the precipitate of the left supernatant Co was detected, and from that of the right one W was detected. Therefore, it can also be inferred that the same reaction as the experiment using the reagent occurred.

6   Conclusions

The recovery method of W and Co in ECM of sintered carbide was investigated. The conclusions obtained are as follows.

- It was shown that W can be recovered in ECM of sintered carbide in the form of C $CaWO_4$ or $WO_3 \cdot H_2O$ by adding Ca $(NO_3)_2$ or HNO_3 into the electrolyte
- It was confirmed that Co can be recovered in the form of Co $(OH)_2$ by adding NaOH to the electrolyte.
- When adding Na_2WO_4 and Co $(NO_3)_2$ to water, it turned out that a purple substance was generated and precipitated. The possibility of recovering Co and W during ECM of sintered carbide was shown.
- Through the ECM experiment of sintered carbide, it can be inferred that the same reaction as the experiment using the reagent occurred.

Acknowledgements. The authors acknowledge that this research was supported by SIP (Cross-ministerial Strategic Innovation Promotion Program) of the Council for Science, Technology, and Innovation (The managing corporation is NEDO.) and Program for Building Regional Innovation Ecosystems (Ministry of Education, Culture, Sports, Science and Technology).

References

1. Maeda, S., Saito, N., Haishi, Y.: Principle and characteristics of electro-chemical machining. Mitsubishi Electric Technical Repot (Mitsubishi Denki Giho) **41**(10), 1267–1279 (1967). in Japanese
2. Kurafuji, H.: Electrochemical machining. J. Inst. Electr. Eng. Jpn. **85-5**(920), 743–747 (1965). in Japanese
3. Masuzawa, T., Kimura, M.: Electrochemical surface finishing of tungsten carbide alloy. Ann. CIRP **40**(1), 199–202 (1991)
4. Walther, B., Schilm, J., Michaelis, A., Lohrengel, M.M.: Electrochemical dissolution of hard metal alloys. Electrochim. Acta **52**, 7732–7737 (2007)
5. Shibuya, N., Ito, Y., Natsu, W.: Electrochemical machining of tungsten carbide alloy micro-pin with NaNO3 solution. Int. J. Precis. Eng. Manuf. **13**(11), 2075–2078 (2012)
6. Choi, S.H., Kim, B.H., Shin, H.S., Chu, C.N.: Analysis of the electrochemical behaviors of WC–Co alloy for micro ECM. J. Mater. Process. Technol. **213**, 621–630 (2013)
7. Mizugai, K., Shibuya, N., Kunieda, M.: Study on electrolyte jet machining of cemented carbide. Int. J. Electr. Mach. (IJEM) **18**, 23–28 (2013)

8. Koyanoa, T., Kunieda, M.: Ultra-short pulse ECM using electrostatic induction feeding method. Proc. CIRP **6**, 390–394 (2013)
9. Natsu, W., Kurahata, D.: Influence of ECM pulse conditions on WC alloy micro-pin fabrication. Proc. CIRP **6**, 401–406 (2013)
10. Schubert, N., Schneider, M., Michaelis, A.: Electrochemical machining of cemented carbides. Int. J. Refract. Metals Hard Mater. **47**, 54–60 (2014)
11. Mogilnikov, V.A., Chmir, M.Y., Timofeev, Y.S., Poluyanov, V.S.: Diamond-ECM grinding of sintered hard alloys of WC-Ni. Proc. CIRP **42**, 143–148 (2016)
12. Miyoshi, K., Kunieda, M.: Fabrication of micro rods of cemented carbide by electrolyte jet turning. Proc. CIRP **42**, 373–378 (2016)
13. Takagi, K., Aoki, H.: Synthesis of calcium tungstate phosphor by solid-state reaction between calcium oxide and tungsten oxide. J. Ind. Chem. **61**(2), 267 (1961)
14. Wells, A.F.: Structural Inorganic Chemistry, 5th edn. Clarendon Press, Oxford (1986). ISBN 0-19-855370-6
15. Montini, T., Gombac, V., Hameed, A., et al.: Synthesis, characterization and photocatalytic performance of transition metal tungstates. Chem. Phys. Lett. **498**(9), 113–119 (2010)

Deuterium Removal Efficiency in Tungsten as a Function of Hydrogen Ion Beam Fluence and Temperature

Mingzhong Zhao[1(✉)], Qilai Zhou[2], Moeko Nakata[2], Akihiro Togari[2],
Fei Sun[2], Yuji Hatano[4], Naoaki Yoshida[5], and Yasuhisa Oya[3]

[1] Graduate School of Science and Technology, Shizuoka University,
Shizuoka, Japan
zhao.mingzhong.17@shizuoka.ac.jp
[2] Faculty of Science, Shizuoka University, Shizuoka, Japan
[3] Graduate School of Integrated Science and Technology, Shizuoka University,
Shizuoka, Japan
[4] Hydrogen Isotope Research Center, Organization for Promotion of Research,
University of Toyama, Toyama, Japan
[5] Research Institute for Applied Mechanics, Kyushu University, Fukuoka, Japan

Abstract. Establishment of effective tritium removal method was one of important issues for the development of fusion reactor from the view of fuel recycle and safe operation. The deuterium (D) removal efficiency in tungsten (W) by energetic hydrogen (H) ions under room temperature and baking under 623 K were studied by thermal desorption spectroscopy (TDS). Iron (Fe) damaged W with various damage level by 6 MeV Fe^{2+} was adopted to simulate neutron irradiation damages. To understand the D removal behavior, the desorption of D_2 was measured in-situ by a quadrupole mass spectrometer (QMS) during H_2^+ implantation and baking. The in-situ results showed that the desorption of D_2 started after H_2^+ implantation and became slowly with the increment of H_2^+ implantation time. After H_2^+ implantation, part of D trapped by dislocation loops, vacancy clusters and voids could be removed by hydrogen isotope exchange. However, the removal efficiency by hydrogen isotope exchange decrease obviously as the presence of irradiation damages. The D trapped by dislocation loops and vacancy clusters can be removed by baking with high efficiency. It is worth to note that the D trapped by voids cannot be removed by baking leading to the lower D removal efficiency for W with high damage level.

Keywords: Tungsten · Hydrogen isotope exchange · Irradiation damages

1 Introduction

Tritium (T) retention in plasma facing material is an important issue for fusion reactors due to the safety limitation of total T retention and T self-sufficiency. As a fusion device under construction, the maximum T retention in the in-vessel components of International Thermonuclear Experimental Reactor (ITER) is set to be 700 g [1, 2]. To

© Springer Nature Switzerland AG 2019
G. Laukaitis (Ed.): INTER-ACADEMIA 2018, LNNS 53, pp. 20–27, 2019.
https://doi.org/10.1007/978-3-319-99834-3_3

prolong the working time of plasma facing component and reduce the operation cost, the development of T removal technique will be necessary for fusion devices. Baking under 513 K and 623 K for the first wall and divertor has been adopted as the T removal technique in ITER [2, 3].

In addition, hydrogen isotope exchange has also been proposed as the T removal method [1]. Various hydrogen isotope exchange experiments using plasma [4–6] and neutral hydrogen (H) gas [7] have been performed in tungsten (W) which is one of the candidate plasma facing materials for fusion reactor. The results show that deuterium (D) retention in W can be removed by subsequently hydrogen (H) exposure. The removal efficiency is associated with H fluence, temperature [8] and irradiation damage level [6, 9]. However, the energy of hydrogen isotope particles ranges from eV to several keV in fusion reactor [10]. There are few reports on the hydrogen isotope exchange behavior in W under the action of energetic hydrogen isotopes.

In the present work, D removal in W by energetic H ions and baking was studied. As W has been selected as the plasma facing material in ITER divertor region [11, 12], the divertor baking temperature of ITER (623 K) was used in this work [2]. To evaluate the effect of irradiation damages on D removal behavior, the present experiments were also performed on iron (Fe) damaged W with different damage level.

2 Experimental

Material used in the present study was polycrystalline W sample cut from W rod which was supplied by Allied Material (A.L.M.T.) Corp. Ltd. The size of sample was 10 mm in diameter and 0.5 mm in thickness. Both side of samples were polished to mirror surface. To release the internal stress, heating treatment under high vacuum ($<10^{-6}$ Pa) up to the temperature of 1173 K for 30 min was performed for all of the samples.

To understand the effect of irradiation damage on the D removal efficiency, Fe^{2+} irradiation with energy of 6 MeV under room temperature was performed by Takasaki Ion Accelerators for Advanced Radiation Application (TIARA). Two different damage levels were used, one is 0.01 dpa and the other is 1 dpa. It was worth to note that 0.01 and 1 dpa were the mean value of dpa throughout the implantation depth. The irradiation depth of 6 MeV Fe^{2+} is up to about 1.8 μm with the peak damage at about 1 μm calculated by Stopping and Range of Ions in Matter (SRIM) with the displacement threshold energy of 50 eV [13, 14]. After Fe^{2+} irradiation, the microstructure of W sample was analyzed by JEM 2000EX Transmission Electron Microscope (TEM) at Kyushu University.

Hydrogen exchange experiment was performed using Triple Ion Implantation system in Shizuoka university as shown in Fig. 1(a) [15]. The D_2^+ implantation with energy of 3 keV was adopted to introduce D into the materials. Subsequently, The H_2^+ implantation with the same energy was used to study the hydrogen exchange behavior. The depth distributions of implanted H and D calculated by SRIM were given in Fig. 1 (b). Both of the D_2^+ and H_2^+ implantation flux were set to be 1.0×10^{18} ions/m^2 s^{-1}. The fluence of D_2^+ implantation was fixed at 1.0×10^{22} ions/m^2. For H_2^+ implantation, the implantation fluence varies from 1×10^{19} to 1×10^{21} ions/m^2. Both of D_2^+ and H_2^+ implantation were performed under room temperature to exclude the effect of temperature on results. In order to study the hydrogen isotopes exchange process, the desorption of D_2 was measured in-situ by a quadrupole mass spectrometer (QMS, MKS Inc Microvision Plus) during H_2^+ implantation.

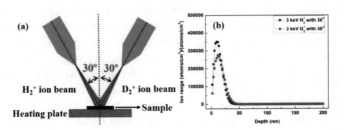

Fig. 1. (a) Schematic view of the H_2^+ and D_2^+ implantation, (b) the ions distribution as a function of depth calculated by SRIM

Baking experiment was performed on the D_2^+ implanted sample under vacuum. The sample was heated to 623 K with the heating rate of 30 K/min. Then the temperature was held at 623 K for 1 h until the D_2 signal decreased to background level.

Finally, the D retained behavior in W was measured by thermal desorption spectroscopy (TDS) with the heating rate of 30 K/min up to 1173 K. Both of Mass 4 and Mass 3, calibrated by D_2 standard leak were monitored by QMS.

3   Results

3.1   Microstructure and D Retention in Damaged Sample

The TDS spectra for undamaged and damaged W after D_2^+ implantation were shown in Fig. 2(a). The D retention was clearly controlled by the damage level. Three desorption peaks at 400, 700 and 830 K, named as Peaks 1, 2 and 3, were observed. According to our previous studies, the Peaks 1, 2 and 3 corresponded to the D trapped by dislocation loops or surface, vacancy clusters and voids, respectively [13, 14, 16].

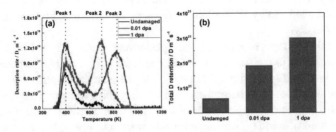

Fig. 2. D retention behavior for undamaged and damaged W. (a) D_2 desorption spectra, (b) the total D retention

For the undamaged sample, the Peak 1 was the major desorption stage, indicating that most of D was trapped by dislocation loops or surface. In the case of 0.01 dpa damaged W, the D_2 desorption was concentrated on Peaks 1 and 2, suggesting that vacancy clusters were formed by Fe^{2+} irradiation. As the damage level has increased to 1 dpa, the major desorption peak became to Peaks 1 and 3. This indicates that dislocation loops and voids work as major D tapping sites for 1 dpa damaged sample. The total D retentions for samples with various damage level were summarized in Fig. 2(b). Due to the existence of irradiation defects, the total amount of D retention increased with damage level.

The various D trapping sites for the damaged W were confirmed by TEM results as shown in Fig. 3. The dislocation loops, black dots in Fig. 3(a), were observed in 0.01 dpa damaged sample. For the higher damage level of 1 dpa, both of dislocation loops and voids were observed. These results indicated that much more stable trapping site for D would be formed and the total D retention in W would increase as the damage concentration increase.

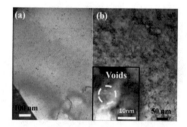

Fig. 3. The microstructure of W damaged by Fe^{2+} ions, (a) 0.01 dpa damaged sample, (b) 1 dpa damaged sample [17]

3.2 Hydrogen Isotope Exchange

The TDS spectra of D_2 released from the samples with various H_2^+ implantation fluence were given in Fig. 4(a). Both of the D_2 trapped by dislocation loops and vacancy clusters reduced after H_2^+ implantation. It was noted that D_2^+ and H_2^+ implantation were performed under room temperature. Hence, the depletion of D in W was caused by hydrogen isotope exchange.

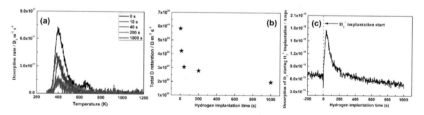

Fig. 4. (a) D_2 desorption spectra after H_2^+ implantation, (b) D retention as a function H_2^+ implantation time, (c) desorption of D_2 during H_2^+ implantation

The evolution of total D retention as a function of H_2^+ implantation time was shown in Fig. 4(b). The total D retention decreased quickly at the initial of H_2^+ implantation, then the depletion of D retention became slowly. To well understand the desorption of D_2 by hydrogen isotope exchange, the in-situ D_2 signal during H_2^+ implantation was given in Fig. 4(c). The in situ results were in good agreement with the total D retention measured by TDS. The desorption of D_2 from W occurred after H_2^+ implantation start. As the increase of H_2^+ implantation time, the D_2 signal decreased to background level gradually.

The hydrogen exchange experiment was also performed on 0.01 and 1 dpa Fe^{2+} damaged samples. As shown in Fig. 5, the D_2 release behavior during H_2^+ implantation was similar among undamaged and Fe^{2+} damaged samples. The release of D_2 was terminated at about 1000 s H_2^+ implantation. According to the results of undamaged sample (Fig. 4), it can be concluded that the D retention in damaged sample decreased quickly after H_2^+ implantation. With the increase of H_2^+ implantation time, the D retention became stable. Hence, 1000 s H_2^+ implantation was adopted to compare the D removal between undamaged and damaged samples.

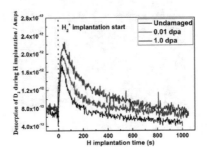

Fig. 5. In-situ D_2 desorption spectra during H_2^+ implantation

As shown in Fig. 6, the TDS spectra of D_2 gave the detailed information about the depletion of D_2 trapped by various damages. It was clear to see that part of D_2 trapped by dislocation loops, vacancy clusters and voids could be removed by subsequent H_2^+ implantation. The D trapped by vacancy clusters and voids was much more difficult to

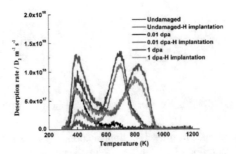

Fig. 6. D_2 TDS spectra of the samples with and without H_2^+ implantation

remove compared with that by dislocation loops. As reported in previous studies, the trapping energy of D by dislocation loops, vacancy clusters and voids corresponded to 0.65, 1.25 and 1.55 eV, respectively [13, 18]. Therefore, higher energy was required to initiate the release of D trapped by vacancy clusters and voids than dislocation loops. This leaded to the D removal efficiency decrease with the increment of irradiation damage level. The D removal efficiency for undamaged sample was about 66% after 1000 s H_2^+ implantation. For the 0.01 dpa damaged sample, the D removal efficiency was about 21%. The D removal efficiency decreased to about 7% for the 1dpa damaged sample.

3.3   Deuterium Retention After Baking

The D retention behavior after baking under 623 K was studied by TDS. The D_2 release spectra during heating up and heating preservation were shown in Fig. 7(a). For the heating up process, the D_2 release behavior was similar with the TDS spectra as shown in Fig. 2(a). After getting into heating preservation, the D_2 signal was gradually reduced with time. After about 1000 s heating preservation, the D_2 signal was dropped to the background level. This implied that there was no further release of D_2 with the increment of baking time. The D retention in W material was measured by TDS after baking experiment. The TDS spectra for undamaged, 0.01 dpa and 1dpa damaged samples were given in Fig. 7(b)–(d) respectively. As a comparison, the TDS spectra of D_2^+ implantation only and after H exchange were also shown. It was clear that D trapped by dislocation loops and vacancy clusters were removed by baking. However, the D trapped by voids was difficult for the removal under the present temperature of

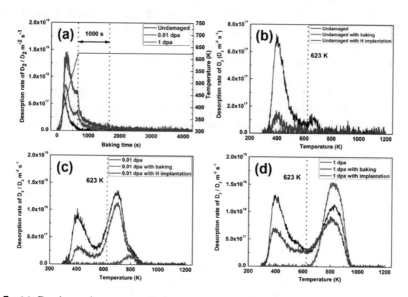

Fig. 7. (a) D_2 desorption spectra during baking experiment. Comparison of D_2 desorption behavior after baking at 623 K and 1000 s H_2^+ implantation, (b) undamaged sample, (c) 0.01 dpa and (d) 1 dpa

623 K. It was reported that voids in W were stable up to the temperature of 1023 K [19, 20]. Some of D desorption from dislocation loops and vacancy clusters would be re-captured by voids during diffusion process in the material. This would lead to the increase of Peak 3 as shown in Fig. 7(d).

The total D retention under various experimental conditions were summarized in Table 1. After baking, the total D retention obviously reduced for the undamaged and 0.01 dpa damaged samples. The existence of voids stably retained large amount of D for the 1 dpa damaged sample after baking. Hence, the D trapped by voids plays a key role in the assessment of T retention.

Table 1. The total D retention (D/m^2) in W under various experiment condition

Experiment	Sample		
	Undamaged	0.01 dpa	1 dpa
D_2^+ implantation only	5.83×10^{20}	1.89×10^{21}	3.01×10^{21}
After H_2^+ implantation	1.96×10^{20}	1.49×10^{21}	2.80×10^{21}
After baking	1.27×10^{20}	2.94×10^{20}	1.88×10^{21}

4 Conclusion

Deuterium removal in W by hydrogen isotope exchange and baking was studied by TDS. To evaluate the removal of D trapped by various irradiation defects, Fe^{2+} irradiated W with various damage level were used. The in-situ measured results indicated that the release of D_2 by H_2^+ implantation or baking gradual finish at about 1000 s under the present experimental condition. After H_2^+ implantation, some of D trapped by dislocation loops, vacancy clusters and voids can be removed. The presence of irradiation damages reduced the remove efficiency obviously. In the case of baking under 623 K, D trapped by dislocation loops and vacancy clusters can be removed. However, the D trapped by voids could not be removed by baking under 623 K. This lead to the low D removal efficiency in the sample with high damage level. Hence, the voids induced by neutron irradiation would play a key role for the assessment of T retention.

Acknowledgement. This study was supported by the NIFS collaboration research.

References

1. Roth, J., Tsitrone, E., Loarer, T., et al.: Tritium inventory in ITER plasma-facing materials and tritium removal procedures. Plasma Phys. Control. Fusion **50**, 103001 (2008)
2. De Temmerman, G., Baldwin, M.J., Anthoine, D., et al.: Efficiency of thermal outgassing for tritium retention measurement and removal in ITER. Nucl. Mater. Energy **12**, 267–272 (2017)

3. Nobuta, Y., Hatano, Y., Torikai, Y., et al.: Effects of baking in deuterium atmosphere on tritium removal from tungsten. Fus. Eng. Des. (2018, in press)
4. Watanabe, T., Kaneko, T., Matsunami, N., et al.: In-situ measurement of deuterium retention in W under plasma exposure. J. Nucl. Mater. **463**, 1049–1052 (2015)
5. Bobyr, N.P., Alimov, V.Kh., Khripunov, B.I., et al.: Influence of helium on hydrogen isotope exchange in tungsten asequential exposures to deuterium and helium–protium plasmas. J. Nucl. Mater. **463**, 1122–1124 (2015)
6. Barton, J.L., Wang, Y.Q., Schwarz-Selinger, T., et al.: Isotope exchange experiments in tungsten with sequential deuterium and protium plasmas in PISCES. J. Nucl. Mater. **438**, S1183–S1186 (2013)
7. Markelj, S., Založnik, A., Schwarz-Selinger, T., et al.: In situ NRA study of hydrogen isotope exchange in self-ion damaged tungsten exposed to neutral atoms. J. Nucl. Mater. **469**, 133–144 (2016)
8. Roth, J., Schwarz-Selinger, T., Alimov, V., et al.: Hydrogen isotope exchange in tungsten: discussion as removal method for tritium. J. Nucl. Mater. **432**, 341–347 (2013)
9. Ogorodnikova, O.V., Markelj, S., Eflimov, V.S., et al.: Deuterium removal from radiation damage in tungsten by isotopic exchange with hydrogen atomic beam. J. Phys.: Conf. Ser. **748**, 012007 (2016)
10. Roth, J., Schmid, K.: Hydrogen in tungsten as plasma-facing material. Phys. Scr. **T145**, 014031 (2011)
11. Pitts, R.A., Carpentier, S., Escourbiac, F., et al.: A full tungsten divertor for ITER: physics issues and design status. J. Nucl. Mater. **438**, S48–S56 (2013)
12. Buzi, L., De Temmerman, G., Huisman, A.E., et al.: Response of tungsten surfaces to helium and hydrogen plasma exposure under ITER relevant steady state and repetitive transient conditions. Nucl. Fusion **57**, 126009 (2017)
13. Fujita, H., Uemura, Y., Sakurada, S., et al.: The damage depth profile effect on hydrogen isotope retention behavior in heavy ion irradiated tungsten. Fus. Eng. Des. **125**, 468–472 (2017)
14. Fujita, H., Yuyama, K., Li, X., et al.: Effect of neutron energy and fluence on deuterium retention behaviour in neutron irradiated tungsten. Phys. Scr. **T167**, 014068 (2016)
15. Oya, Y., Suzuki, S., Wang, W., et al.: Correlation between deuterium retention and microstructure change for tungsten under triple ion implantation. Phys. Scr. **T138**, 014051 (2009)
16. Ryabtsev, S., Gasparyan, Y., Zibrov, M.: Deuterium thermal desorption from vacancy clusters in tungsten. Nucl. Instrum. Methods Phys. Res. B **382**, 101–104 (2016)
17. Oya, Y., Azuma, K., Togari, A., et al.: Interaction of hydrogen isotopes with radiation damaged tungsten. Advances in Intelligent Systems and Computing **660**, 41–49 (2018)
18. Oya, Y., Hatano, Y., Shimada, M., et al.: Recent progress of hydrogen isotope behavior studies for neutron or heavy ion damaged W. Fusion Eng. Des. **113**, 211–215 (2016)
19. Nambissan, P.M.G., Sen, P.: Positron annihilation study of the annealing behaviour of alpha induced defects in tungsten. Radiat. Eff. Defects Solids **124**, 215–221 (1992)
20. Ogorodnikova, O.V., Sugiyama, K.: Effect of radiation-induced damage on deuterium retention in tungsten, tungsten coatings and Eurofer. J. Nucl. Mater. **442**, 518–527 (2013)

Formation and Investigation of Doped Cerium Oxide Thin Films Formed Using E-Beam Deposition Technique

Nursultan Kainbayev[1,2(✉)], Mantas Sriubas[1], Zivile Rutkuniene[1],
Kristina Bockute[1], Saltanat Bolegenova[2], and Giedrius Laukaitis[1]

[1] Physics Department, Kaunas University of Technology, Studentu str. 50,
51368 Kaunas, Lithuania
nursultan.kainbayev@ktu.edu
[2] Department of Thermal Physics and Technical Physics, Al-Farabi Kazakh
National University, 71 al-Farabi Ave, Almaty, Kazakhstan

Abstract. The investigation of new functional materials (ceramics) based on cerium (IV) oxides is a promising field of scientific research. A wide application in the industry received composite materials based on CeO_2–Gd_2O_3 and CeO_2–Sm_2O_3.

Thin ceramic films were formed on the basis of CeO_2 with 10 mol% Gd_2O_3 (GDC10), CeO_2 with 20 mol% Gd_2O_3 (GDC20), CeO_2 with 15 mol% Sm_2O_3 (SDC15), CeO_2 with 20 mol% Sm_2O_3 (SDC20) using e-beam technique in this work. The deposition rate and temperature of the substrate had influence on the formed doped cerium oxide GDC10, GDC20, SDC15, SDC20 thin films structure. Sm and Gd doped cerium oxide thin films were deposited on SiO_2, Alloy 600 (Fe-Ni-Cr), Si (111), Si (100) and Al_2O_3 substrates. Investigations of the formed thin films were carried out using a Scanning electron microscope (SEM), Electron dispersive spectroscopy (EDS), X-ray diffraction (XRD), and Raman spectroscopy. It has been established that the cerium oxide based ceramic retains the crystalline structure, regardless of the concentration of the dopant and used substrate type. The most dominant crystallographic orientation of formed thin films was cubic (111). Raman spectroscopy measurements showed the peak (465 cm^{-1}) of pure ceria corresponding to F_{2g} vibrational mode. First-order peaks, inherent to cerium oxide, were shifted to a region of lower wavenumbers and depend on dopant concentration. The peaks for all formed thin films were similar to each other in form but the position, half width and their intensity varied depending on the dopant concentration. Raman peaks position at 550 cm^{-1} and 600 cm^{-1} could be explained as change of oxygen vacancy amount due to the cerium transition between oxidized and reduced forms.

Keywords: Ceramics · Cerium oxide thin films
Doped cerium oxide thin films · Raman spectroscopy

© Springer Nature Switzerland AG 2019
G. Laukaitis (Ed.): INTER-ACADEMIA 2018, LNNS 53, pp. 28–34, 2019.
https://doi.org/10.1007/978-3-319-99834-3_4

1 Introduction

The creation of new functional materials based on oxides (catalysts, counter electrodes of electrochromic devices, solid oxide fuel cells), including cerium (IV) oxide, is a promising field of scientific research corresponding to priority areas of Lithuania's scientific and technological development "Technologies for the production and processing of functional nanomaterials". The composite materials of cerium oxide doped with gadolinium and samarium are widely used in industry. These materials could have cerium transition between the oxidized and reduced forms [1]. Gd (IV) and Sm (IV) oxides influence this equilibrium, and also increase the thermal stability and transmittance in the visible region of the ceria (IV) oxide spectrum. The known physical and chemical, combined methods for the production of oxide CeO_2–CeO_2–Gd_2O_3 and CeO_2–Sm_2O_3 in thin-film states are mainly energy-intensive and require the use of expensive equipment. In our case, were have chosen the method of electron beam evaporation (E-beam deposition). In the method of electron beam evaporation, changes in the deposition rate, pressure and temperature of the substrate affect the properties of oxide films. By changing the application parameters, were could obtain thin films with a homogeneous structure and a homogeneous shape without any defects and cracks [2, 3]. A fuel cell is a device that converts chemical energy into electrical energy. There are different types of fuel cells designed for a specific application, the most important difference between them is the kind of electrolyte they use. This study was carried out using a solid oxide electrolyte, which is a cerium oxide with (Gd_2O_3 and Sm_2O_3) dopant. The properties of a thin film electrolyte in many respects define operational characteristics of the fuel cell. High ionic conductivity, mechanical strength thermal resistance and gas tightness are the main requirements to the electrolyte ensure stable and long-term operation of fuel cells with high electrical performance, preventing the loss of fuel. The purpose of the research is to form a thin film on the basis of cerium oxide (CeO_2) with dopants Sm and Gd using e-beam evaporation technique, to investigate the microstructure and Raman spectra study are the purpose of the research. The early theories, on the basis of which the search for an optimum dopant for CeO_2 was carried out, were generally based on consideration of deformation of a lattice in general and did not take into account local microstructure of a solid electrolyte. With development of new methods of the analysis there was an opportunity to find the additional parameters influencing the properties of solid solution which are not connected with technological features of their production. The method of Raman spectroscopy enabled us to obtain additional information on oxygen vacancies in the investigated solid solutions.

2 Experimental

In the present study Gadolinium and Samarium doped ceria were formed with e-beam physical vapor deposition system "Kurt J. Lesker EB-PVD 75", using 0.2 nm/s ÷ 1.6 nm/s deposition rate and temperatures from 50 °C to 600 °C of SiO_2, Alloy 600 (Fe-Ni-Cr), Si (111), Si (100) and Al_2O_3 substrates. The $Sm_{0.1}Ce_{0.9}O_{2-\delta}$, $Sm_{0.2}Ce_{0.8}O_{2-\delta}$ and $Gd_{0.1}Ce_{0.9}O_{1.95}$, $Gd_{0.2}Ce_{0.8}O_{1.95}$ powders (Nexceris, LLC, Fuelcellmaterials,

USA) was used as evaporating material. The pellets were placed into crucible and vacuum chamber was depressurised up to $2 \cdot 10^{-9}$ bar. After that, the substrates were treated with Ar+ ion plasma (10 min) and preheated up to working temperature. Thickness and deposition rate were controlled with INFICON crystal sensor. Formed thin films ceramic base on Cerium oxides structures and morphology have been studied by X-ray diffraction (XRD) "Bruker D8 Discover" at 2Θ angle in a 20°–70° range using Cu Kα (λ = 0.154059 nm) radiation, 0.01° step, and Lynx eye PSD detector. EVA Search – Match software and PDF-2 database were used to identify diffraction peaks. The crystallite size was calculated using Scherrer's equation. Scanning electron microscopy (SEM) "Hitachi S-3400N", was used to obtain surface topography images and cross-section images Elemental composition was controlled using energy-dispersive X-ray spectroscopy "BrukerXFlash QUAD 5040" (EDS). Raman spectra were recorded at the room temperature, using a Confocal Raman spectrometer Solver Spectrum (NT-MDT). The excitation source was the 532 nm, diameter of laser spot was 2 microns and diffraction grating 1800/500 providing spectral resolution of 1 cm^{-1}. Laser output was 14 mW. Fitting procedure were carried out by the data of Lorentz line shape using a peak fit in the OriginPro software.

3   Results and Discussions

Various concentration of doped gadolinium and samarium were used wishing to determine how concentration of impurities might affect to the microstructure and ionic conductivity of thin films, CeO_2 with 10 mol% Gd_2O_3 (GDC10), CeO_2 with 20 mol% Gd_2O_3 (GDC20), CeO_2 with 15 mol% Sm_2O_3 (SDC15), CeO_2 with 20 mol% Sm_2O_3 (SDC20). XRD measurement of thin films was carried out through the program (DIFFRACTplus EVA) and shown in Fig. 1.

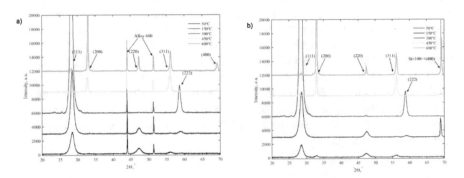

Fig. 1. XRD patterns of SDC20 deposition rate of 1.6 nm/s thin ceramic deposited on a) Alloy 600 and b) Si (100)

The crystallite size was calculated using the program (TOPAS). The results were analyzed by comparison of deposition rate and substrate temperature and detected that, all thin films (GDC10, GDC20, SDC15, SDC20) based on cerium oxide have a single-phase and fluorite structure with space group Fm3m.

X-ray diffraction patterns of SDC20 thin films have characteristic peaks, corresponding to crystallographic orientations (111), (200), (220), (311), (222), and (400). For the formed thin films, preferential orientations are (111) and (222). However, investigated ceramic thin films do not retain the same structure and orientation, since the deposition parameters were different, they change into (200) or (311) (Figs. 1, 2), i.e., it was observed that the change appear at high substrate temperatures (450 °C and 600 °C) and at high deposition rates (1.2 nm/s and 1.6 nm/s). Analysis of thin films of identical components at different deposition parameters might give different structure and orientation.

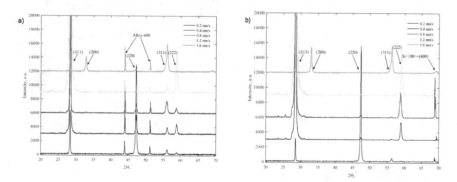

Fig. 2. XRD patterns of SDC20 deposition rate of 0.2 nm–1.6 nm thin ceramic deposited on (a) Alloy 600 and (b) Si (100), substrates temperature $T_{sub} = 450$ °C

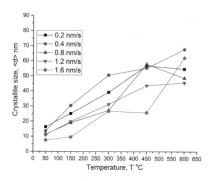

Fig. 3. Temperature dependence of crystallite size at different deposition rate by 0.2 nm/s ÷ 1.6 nm/s thin ceramics deposited on Alloy 600

SDC20 dominant crystallographic orientation (222) the intensity of the peak increases and in some cases, are disappear. This orientation may be associated with deposition energy of the particles, which can migrate into larger clusters changing the dominant peak intensity. In the analysis of XRD, the more intense peaks of a narrow line width shown in Figs. 2 and 3 may indicate an increase in the grain and crystallite size with an admixture of Sm^{3+}.

The calculation of the crystallite size, were prove the statement that by increasing the substrate temperature the crystallite size increases (6.2 nm ÷ 80.1 nm) (Fig. 3). The deposition rate also, have influence on crystallite size. The crystallite size shows almost the same value with increasing the deposition rate if temperature of the substrates, are under 300 °C and crystallite size decreases with increasing deposition rate if substrates temperatures are 450 °C and 600 °C.

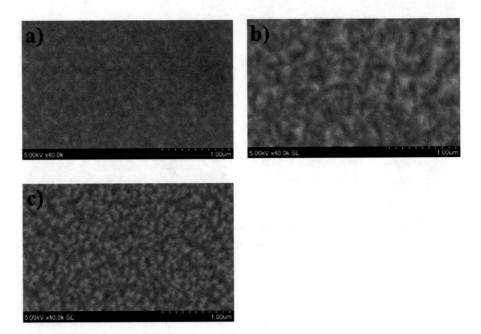

Fig. 4. SEM picture of SDC20 thin ceramic films deposited on optical quartz substrates (SiO_2) using (a) 0.2 nm/s, 50 °C (b) 0.8 nm/s, 300 °C (c) 0.8 nm/s, 600 °C

Depending on the temperature of the substrate, the grain size of the thin film changes as shown in Fig. 4. When the temperature of the substrates increases, the grain sizes increase, but this process occurs up to a certain temperature, in our case to 450 °C, and after that at a temperature of 600 °C, were notice a decrease in the grain size. This is due to the increased influence of the growth rate at high temperatures. As shown in Fig. 4, all the thin films are uniform and do not have any defect or cracks. The obtained results are in good agreement with other authors results [5, 6].

Main peak of pure ceria presents on 465 cm^{-1} on F_{2g} vibrational mode. In this mode can be move only oxygen atoms around the cerium ions Ce^{4+} [7]. As were know in the fluorite lattice only oxygen atoms can be mobile, and the frequency of this mode should not depend on the mass of the cation. It is for this reason that the characteristics of the peaks are very sensitive to disorder induced in the oxygen ion sublattice of the oxide, in particular, the peak position and the width of the peaks [8]. Doping the crystal lattice of cerium with rare-earth Gd and Sm elements, were got a defect in the crystal lattice that affected the environment of oxygen around the metal ions. As a result, were received changes in the position of peaks and width of peaks in the Raman spectra. Considering thin films of cerium oxide doped with Sm and Gd, it is seen that the F_{2g} modes, in particular, first-order peaks inherent to cerium oxide, shifted to the region of lower wavenumbers (Fig. 5).

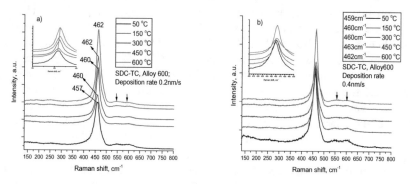

Fig. 5. Raman spectra of Sm doped ceria (SDC20) on Alloy600 substrate at different temperatures by 50 °C to 600 °C and deposition rate: (a) 0.2 nm/s, (b) 0.4 nm/s

The Raman peak position shift consists from 2 cm^{-1} to 9 cm^{-1} of a low-frequency wave (Fig. 5). The peaks for all the samples were similar to each other in form but the wave numbers and their intensity varied depending on the studied sample. The observed intense bands for doped cerium oxides are due to the Raman regime F_{2g} of fluorite dioxide belonging to the space group Fm3m. The shift of peak position to the region of lower wavenumbers shows that the F_{2g} mode corresponds to the symmetric vibrations of oxygen ions around Ce^{4+} ions in octahedral CeO_8 [8]. The peaks for all the samples were similar to each other in a form. Peak position due to the F_{2g} symmetry and shift in the F_{2g} mode is explained by the change in M-O vibration frequency after the doping with Gd and Sm rare-earth elements [9]. Some additional broad peaks expressed in all samples at 530–550 and around 600 cm^{-1} might be explained by presence of oxygen vacancies [9] are created as charge compensation defects induced by the introducing of other metal cations into the crystal lattice of cerium. Peaks at around 600 cm^{-1} could be explained by non-stoichiometry oxygen vacancy in cerium oxide. The appearance of oxygen vacancy in Gd and Sm doped ceria, could be explained by reaction:

$$\text{Oxygen vacancy model}: \quad RE_2O_3 + 2CeO_2 \rightarrow 2RE'_{Ce} + 3V_O^x + V_{\ddot{O}} \qquad (1)$$

$$\text{Interstitial model}: \quad Gd_2O_3 + 2CeO_2 \rightarrow 2Gd'_{Ce} + 3V_O^x + V_{\ddot{O}} \qquad (2)$$

$$Sm_2O_3 + 2CeO_2 \rightarrow 2Sm'_{Ce} + 3V_O^x + V_{\ddot{O}} \tag{3}$$

In the Raman study, peak positions are determined by fitting the data to the Lorentz line shape using a peak fit.

4   Conclusions

SDC20 thin films have a single-phase and fluorite structure with space group Fm3m peaks corresponding to crystallographic orientations (111), (200), (220), (311), (222), and (400). The crystallite size of formed SDC20 thin films increases with increasing temperature of the (Alloy 600 and SiO_2) substrates, the smallest crystallite size is d = 6.2 nm when the temperature of substrate is T_{sub}– 50 °C and biggest crystallite size d = 80.1 nm when T_{sub}– 450 °C). SEM analysis of SDC20 thin films shows that the substrate temperature has the influence on grain size of the formed thin films, the grain size increases till 450 °C substrate temperature and starts to decrease with further increase of the substrate temperature till 600 °C. The main Raman peak of pure ceria is presents at 465 cm^{-1} and corresponds to the F_{2g} vibrational mode. The peaks for all the samples were similar to each other in form, but the wave numbers and their intensity varied depending on the investigated sample. The Raman peak (465 cm^{-1} corresponding to pure ceria) shifts to the lower side from 1 cm^{-1} to 9 cm^{-1} for the formed (GDC10, GDC20, SDC15, SDC20) thin films. That could be influenced by an increase of oxygen vacancies. The most significant shift, 456 cm^{-1} is registered for Sm doped ceria (SDC 15) thin films deposited on SiO_2 and Alloy 600. Raman peaks position at 530–550 cm^{-1} and around 600 cm^{-1} might be explained as change of oxygen vacancy amount due to the cerium transition between oxidized and reduced forms of $Ce^{3+} \rightleftarrows Ce^{4+}$.

References

1. Anwar, M.S., Kumar, S., Arshi, N., Ahmed, F., Lee, C., Koo, B.H.: J. Alloy. Compd. **509**, 4525–4529 (2011)
2. Laukaitis, G., Virbukas, D.: Solid State Ion. **247–248**, 41–47 (2013)
3. Virbukas, D., Laukaitis, G.: Solid State Ion. **302**, 107–112 (2017)
4. Mansilla, C.: Solid State Sci. **11**(8), 1456–1464 (2009)
5. Anwar, M.S.: Curr. Appl. Phys. **11**(Suppl. 1), S301–S304 (2011)
6. Vinodkumar, T., Rao, B.G., Reddy, B.M.: Catal. Today **253**, 57–64 (2015)
7. Salnikov, V.V., Pikalova, E.: Phys. Solid State **57**(10), 1944–1952 (2015)
8. Anjaneya, K.C., Nayaka, G.P., Manjanna, J., Govindaraj, G., Ganesha, K.N.: J. Alloys Compd. **585**, 594–601 (2014)
9. Anjaneya, K.C., Nayaka, G.P., Manjanna, J., Govindaraj, G., Ganesha, K.N.: J. Alloys Compd. **578**, 53–59 (2013)

Two-Photon Fluorescent Microscopy for 3D Dopant Imaging in Wide Bandgap Semiconductors

Amin Al-Tabich[1,2], Wataru Inami[1], Yoshimasa Kawata[1],
and Ryszard Jablonski[2(✉)]

[1] Graduate School of Engineering, Shizuoka University, 3-5-1, Johoku, Naka,
Hamamatsu 432-8561, Japan
[2] Faculty of Mechatronics, Warsaw University of Technology, Sw. A. Boboli 8,
02-525 Warsaw, Poland
r.jablonski@mchtr.pw.edu.pl

Abstract. This paper presents a method of three-dimensional imaging of dopants in wide band gap semiconductors by two-photon fluorescence microscopy. Tightly focused light beam radiated by two titanium-doped sapphire laser is used to obtain two-photon excitation of selected area of the semiconductor sample. Photoluminescence intensity of a specific spectral range is selected by optical band pass filters and measured by photomultiplier tube. Reconstruction of specimen image is done by scanning the volume of interest by piezoelectric positioning stage and measuring the spectrally resolved photoluminescence intensity at each point. The developed two-photon microscope was able to image the whole doped area, which was beyond capabilities of market-available confocal microscope. Additionally, the study found, that the doped volume is immersed below the surface of the semiconductor material, which was impossible to image with confocal microscope. Additionally, the study found, that the doped volume is immersed below the surface of the semiconductor material, which was impossible to image with confocal microscope.

Keywords: Two-photon microscopy · Spectroscopy · Intrinsic defects
Semiconductors

1 Introduction

In many applications the uniformity of the semiconductor crystal is of the highest importance because the internal defects and impurities of a crystal effect the properties of the material, such as light emission efficiency, ferromagnetism, charge carrier mobility, and lifetime. The main advantages of two-photon fluorescence microscopy over standard fluorescence microscopy are: z-axis discrimination thanks to the need of precise focusing on the specimen, reduced harm to the sample, reduced photobleaching effect, deeper sample penetration thanks to longer wavelength and easiness in filtering excitation light from emission light because of wide gap between their wavelengths. As the main disadvantage reduced signal to noise ratio needs to be mentioned. The goal of our research is to widen the spectrum of applications of this technique to imaging and

© Springer Nature Switzerland AG 2019
G. Laukaitis (Ed.): INTER-ACADEMIA 2018, LNNS 53, pp. 35–42, 2019.
https://doi.org/10.1007/978-3-319-99834-3_5

quantifying a selectively doped area of wide band gap semiconductors. In this paper we demonstrate this ability on the example of Zinc Selenide (ZnSe) doped with Gallium (Ga) by the use of focused ion beam (FIB) [1, 2].

2   Experimental Setup

Two-photon fluorescence microscope setup is very similar to classic fluorescence microscope [3–6, 7]. The main change is the need of filtering the excitation light with shortpass filters, instead of longpass. It is also much easier to match the filters, since the

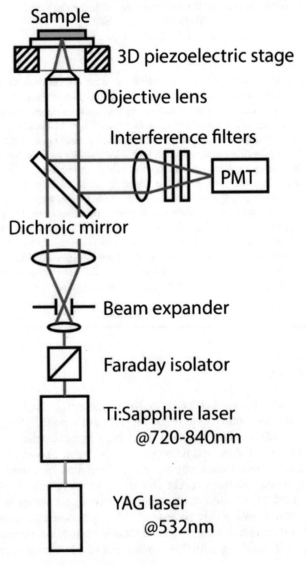

Fig. 1. Experimental setup block diagram

difference in wavelength between excitation and emission light is much more significant than the one resulting only from Stokes shift, observed in classic fluorescence
microscopy. Overlapping of excitation and emitted light spectrum is virtually impossible in the case of two-photon fluorescence. The sample is excited with the beam of
light generated by a Ti:sapphire laser which in used setup can be tuned in the range of
720–840 nm. Additionally, light beam is collimated by a beam expander to fully utilize
aperture of an objective lens and achieve the smallest possible spot size. Reflected
excitation light is filtered out with the use of dichroic mirror and two shortpass filters
and for further attenuation. Such strict filtering leads to approximately 99.995%
extinction of the excitation beam, which might be crucial to achieve high contrast in
this kind of microscope, since the intensity of the light emitted in two-photon process
may be thousands of times lower than for single-photon process, depending on the used
sample. Emitted light is reflected by dichroic mirror and freely passes the filters. It is
then absorbed by a photomultiplier tube PMT. To obtain 2D and 3D images, 3-axis
piezo stage is used to scan the specimen in the focal plane of the objective lens (Fig. 1).

3   Imaging of the Doped Area

To evaluate the abilities of the developed system, the ZnSe monocrystals was studied.
First step was measuring the ZnSe emission and absorption spectra (Fig. 2). Absorption
edge maximum of the studied sample was at 405 nm. As can be noticed from the
emission spectrum, the intensity of emission at liquid nitrogen temperature increased
almost six-fold. Substantial blue-shift was also observed. We can also notice that the
single peak visible in room temperature consists of two overlapping peaks, which can
only be distinguished at the low-temperature spectrum. This second peak can be
attributed to the donor-acceptor pair (DAP) recombination.

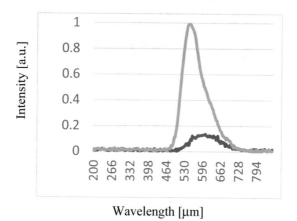

Fig. 2. Emission spectra of the ZnSe monocrystals taken at 300 K (red line) and 77 K
(blue line)

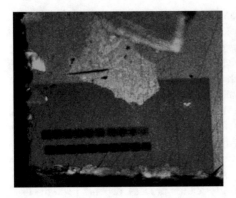

Fig. 3. Confocal fluorescence microscopy image of the processed Ga:ZnSe monocrystal. Resolution: 4096 x 4096 pixels (left). One photon image of Gallium doped ZnSe after 1 h annealing process at 500 °C (right)

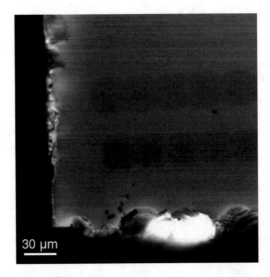

Fig. 4. Two-photon image of the processed surface

The sample was processed by the FIB to introduce Ga dopants into its volume. Various irradiation parameters, current varying from 0.1 to 1 nA and voltage equal 15 kV and 10 kV was tested to study the difference in the depth and nature of the doped area. Confocal image of the surface of processed crystal is shown on Fig. 3 (left).

As can be noticed, not only the directly processed area is distinguishable on the confocal image, but also the square area that was affected by a very low current used for imaging. The ionization pattern was observable throughout the whole volume imaged with the confocal microscopy. Therefore, this method has no ability to determine the depth of ionization. The ion implementation process leads to disintegration of the crystal lattice. The doped ZnSe crystalline structure was recovered by annealing in gaseous nitrogen flow to prevent the oxidation for 1 h at 500 °C. The change in one-photon irradiation response is presented on Fig. 3 (right). The doped area recovered, and the photoluminescence additionally increased, compared to the undoped area. Furthermore, it can be noticed that the area irradiated with low-energy gallium ions during imaging in FIB system, that previously formed a characteristic rectangular area around the processed part, completely disappeared after annealing. This is due to the diffusion of dopants throughout the volume of the sample. Moreover, regarding the scratches and mechanical defects, an overall quality of the surface increased.

The two-photon image of the surface at the doped area is shown in Fig. 4. Comparing it with a confocal image (Fig. 3, left) reveals that the doped area excited by the means of two-photon process exhibits decrease in luminescence intensity, while single-photon case provided an opposite effect. There are multiple possible explanations to that phenomenon, which are not a part of this study. Despite the ambiguity of the photoluminescence decrease, it does not diminish the ability to differentiate the doped volume. Inversely to the confocal images, the density of dopants correlates with the decrease of the intensity and can be estimated with the same analytical methods.

A cross-section two-photon image is shown in Fig. 5. The cross-section was imaged along the area doped with acceleration voltage of 15 kV. The red line on the right hand-side figure marks the exact imaging path. Annealing resulted in increased sensitivity of the sample. So far, the two-photon imaging did not leave any visible damage on the samples. However ZnSe after annealing is significantly affected by the radiation. Both, marks of plane scanning, and cross-sectional scanning are clearly visible.

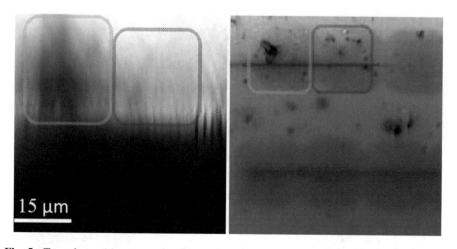

Fig. 5. Two-photon (a) cross section (b) top-down image of the doped area. Cross-section image is take along the red marker

Figure 6 shows a two-photon image of the doped area at different depths. The area was doped from a vertical direction, therefore the shape of the doping does not change or shift. Red oval visible at the image of the surface and 21.36 μm deep marks the defect inflicted by the two-photon imaging at a depth below 10.68 μm inside the sample, which was not visible at the surface. The damage inflicted by the cross-sectional scans was visible through all the volume of the imaging.

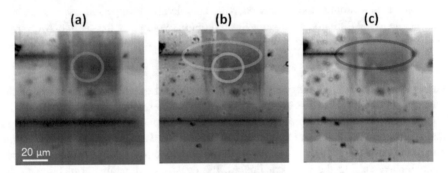

Fig. 6. Two-photon image of the doped ZnSe monocrystal at different depths of focusing. Markers indicate the areas with clear change of the image

Figure 5 can provide an information about the depth of the doping. Blue and purple square markings enclose the doped volume, differentiated by the decrease of the photoluminescence. The blue marking encircles a volume doped with a dose of 1.0 nC/μm^2. The purple marking encloses a volume doped with 0.9 nC/μm^2. To better visualize the intensity decrease in the doped area, an intensity topography of the cross-section image is presented in Fig. 7.

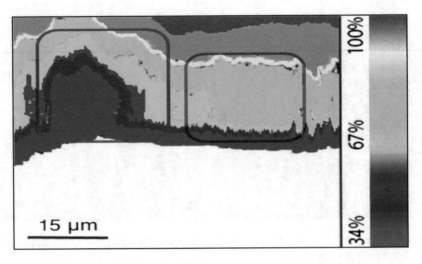

Fig. 7. Intensity distribution of ZnSe monocrystal in the doped volume

The depth of the doped area for the 1.0 nC/µm² implemented with acceleration voltage of 15 kV and 10 kV was calculated as a difference of relative average intensity of the doped area in comparison with the average intensity of the undoped area at the same depth of the semiconductor crystal, to remove the effect of changing photoluminescence with the depth of imaging. The result of this calculation is shown in Fig. 8.

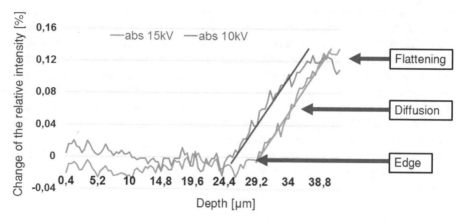

Fig. 8. Relative intensity between doped and not processed volume at given depth of focusing, for 10 kV and 15 kV acceleration voltage of dopant implantation

The presented result indicates that the depth for strong doping is dependent on the acceleration voltage. For the 15 kV case, the doping started to dilute at 73 µm of depth.

For 10 kV, this depth equaled 64 µm. Below this values, the effect of dopant diffusion can be observed as a slope, indicating a gradual increase of the photoluminescence intensity, back to the level of undoped parts of the semiconductor. For both cases, after about 30 µm flattening of the curve can be noticed, which can be attributed to the complete recovery of the original crystalline structure.

4   Conclusion

In the paper a method of three-dimensional imaging of dopants in wide band gap semiconductors by two-photon fluorescence microscopy is presented.

Tightly focused titanium-doped sapphire laser light beam is used to obtain two-photon excitation of selected area of the semiconductor sample. Optical band pass filters can select a specific spectral range of photoluminescence intensity.

Reconstruction of specimen image is done by scanning the volume of interest by piezoelectric positioning stage and measuring the spectrally resolved photoluminescence intensity at each point by photomultiplier tube. Specific spectral range of photoluminescence intensity is selected by means of optical band pass filters.

It was proven, that two-photon fluorescence is a technique capable of three-dimensional imaging of doped area in the wide band gap semiconductors. Imaging abilities were compared with the confocal fluorescence microscopy. The developed two-photon microscope was able to image the whole doped area, which was beyond capabilities of market-available confocal microscope. Additionally, the study found, that the doped volume is immersed below the surface of the semiconductor material, which was impossible to image with confocal microscope. We believe that currently it is the only technique with such possibilities and further investigation might lead to development of industry-class microscope capable of testing the distribution of dopants in wide variety of samples. Such device might have a tremendous impact on the manufacturing of semiconductor devices by giving access to precise engineering in axial dimension with ease. A step forward from the currently available techniques which allow only for rough estimation of the structure in axial direction. This method can be used for quality control of preprocessed semiconductors by detecting mechanical impurities as well as determining the content of intrinsic defects at any stage of production of semiconductor devices. Other methods of volume defect identification, like XRD, provide information only about the existence of various defects, without the ability to image them. Imaging of these defects will help in gaining a better understanding of their source and propagation and might lead to significant improvements in suppressing the formation of impurities during the processing of semiconductors.

References

1. Noor, A.S.M., et al.: Two-photon excited luminescence spectral distribution observation in wide-gap semiconductor crystals. Appl. Phys. Lett. **92**(16), 161106 (2008)
2. Al-Tabich, A., et al.: 3D imaging of intrinsic crystalline defects in zinc oxide by spectrally resolved two-photon fluorescence microscopy. Appl. Phys. Lett. **110**(22), 221106 (2017)
3. Göppert-Mayer, M.: Über Elementarakte mit zwei Quantensprüngen. Ann. Phys. **401**(3), 273–294 (1931)
4. Kaiser, W., Garrett, C.G.B.: Two-photon excitation in $CaF_2;Eu^{2+}$. Phys. Rev. Lett. **7**(6), 229–231 (1961)
5. Davidovits, P.: Scanning laser microscope. Nature **223**, 831 (1969)
6. Denk, W., Strickler, J.H., Webb, W.W.: Two-photon laser scanning fluorescence microscopy. Science **248**(4951), 73–76 (1990)

Synthesis of BiFeO$_3$-Powders by Sol-Gel Process

Sergei A. Khakhomov[1(✉)], Vladimir E. Gaishun[1],
Dmitry L. Kovalenko[1], Alina V. Semchenko[1], Vitali V. Sidsky[1],
Wieslaw Strek[2], Dariusz Hreniak[2], Anna Lukowiak[2],
Natalya S. Kovalchuk[3], Alyaxandr N. Pyatlitski[3],
Vitaliy A. Solodukha[3], and Dmitry V. Karpinsky[4]

[1] F. Skorina Gomel State University, Sovetskaya 104, 246019 Gomel, Belarus
khakh@gsu.by
[2] Institute of Low Temperature and Structures Research PAS, 2 Okolna St.,
Wroclaw, Poland
[3] JSC "INTEGRAL", Korjenevsky Str., 12, 220108 Minsk, Belarus
[4] Scientific-Practical Materials Research Centre of National Academy
of Sciences of Belarus, Minsk, Belarus

Abstract. The present work aims to design and study novel functional materials with multiferroic properties required in electric applications, such as magnetic and magnetoresistive sensors, actuators, microwave electronic devices, phase shifters, mechanical actuators etc. Complex oxides BiFeO$_3$ for analysis of its magnetic properties were synthesized by sol-gel method as powders. The size, shape and degree of crystallinity of the nanoparticles formed by sol-gel method can be controlled by varying the temperature and the ratio of the concentrations of the initial reactants and the stabilizer. To stop the growth of particles in all cases, it is usually enough to cool quickly the reaction mixture. To isolate the nanoparticles, the precipitating solvent is added, which mixes with the reaction system, but poorly dissolves the "protective shells" of the nanoparticles and, therefore, destabilizes the suspension. As a result, the nanoparticles precipitate as powder, which can be separated by centrifugation. The sol-gel method makes it possible to obtain practically monodisperse nanoparticles of various metals oxides.

Keywords: Sol-gel · Film · Powder · Ferromagnets

1 Introduction

Multiferroics have been known as materials exhibiting ferromagnetic and ferroelectric properties at the same time, which have exhibited interesting physical properties as well as a possibility of practical applications. The rhombohedrally distorted simple perovskite structure of BiFeO$_3$ is one of the representative multiferroic materials and has been much interested due to the antiferromagnetic behavior with a relatively high Neel temperature and the ferroelectric behavior with a high Curie temperature. Multiferroic materials, owing to the coexistence of ferroelectricity, ferromagnetism and even

© Springer Nature Switzerland AG 2019
G. Laukaitis (Ed.): INTER-ACADEMIA 2018, LNNS 53, pp. 43–48, 2019.
https://doi.org/10.1007/978-3-319-99834-3_6

ferroelasticity in the same phase, have shown promising applications in nonvolatile information storages, spintronic devices and magnetoelectric sensors. Among the multiferroic materials studied so far, $BiFeO_3$ (BFO) is known to have a rhombohedrally distorted perovskite structure described by space group R3c. It has two order parameters at room temperature: a ferroelectric ordering with high Curie temperature T_C of 1103 K, and long range antiferromagnetic ordering of the G-type with a magnetic transition temperature T_N of 643 K. As the only one single phase multiferroic material which simultaneously possesses ferroelectric and magnetic properties at room temperature, BFO has been one of the most interesting materials studied. At present, the ceramics of BFO have been extensively investigated. Although rhombohedral $BiFeO_3$ (BFO R-phase) has been studied extensively since first discovery in 1960s, electrical properties of the pure BFO R-phase have been rarely reported due to its high conductivity, which may originated from uncertain oxygen stoichiometry, high defect density and poor sample quality [1]. In order to understand the properties of multiferroic BFO, it is very important that the fabrication of pure BFO phase should be established. If temperature and oxygen partial pressure were not controlled accurately during crystallization of the BFO R-phases, the kinetics of phase formation always lead to other impurity phases in Bi-Fe-O system such as $Bi_2Fe_4O_9$, Bi_2O_{3-d} and $Bi_{46}Fe_2O_{72}$.

2   Experimental

Wet chemical methods are promising routes to prepare fine and homogeneous powder. Various wet chemical methods such as hydrothermal, co-precipitation, combustion synthesis, molten-salt method, thermal decomposition, and sol-gel process have been developed and designed to prepare pure $BiFeO_3$ nanopowder. Recently, acid-assisted gel strategy has been proved to be an effective way to synthesize metastable $BiFeO_3$ nanopowder. Pure $BiFeO_3$ phase could be obtained by leaching out the minor Bi_2O_3 phase using diluted nitric acid. Pure $BiFeO_3$ powder can be directly synthesized through the acetic acid-assisted or the tartaric acid-assisted sol-gel method. However, $BiFeO_3$ powder synthesized by the organic acid-assisted sol-gel method might have relatively low phase purity resulting from easy formation of bismuth-rich phase during calcining. Therefore, mineral acid should be considered as an adjuvant to prepare $BiFeO_3$ nanopowder.

Two versions of the sol-gel method [2, 3] were used to synthesize the $BiFeO_3$ powders.

1. $Fe_xBi_yO_z$-citrate-based powder was synthesized using citric acid, ethylenediamine and nitric acid salts of Fe and Bi. Ethylene glycol was used as a solvent. In the beginning, the nitric acid salts of Fe and Bi were dissolved in ethylene glycol without addition of water. Then, citric acid was added to form Fe and Bi citrate. After that, the pH of the solution was adjusted to a value of 7–8 by neutralizing excess citric acid with ethylenediamine. The last neutralization step should be

carried out with constant stirring, dropwise adding ethylenediamine and waiting for a constant pH to be established before the next drop is added. After homogenization of the resulting solution, ethylene glycol was added thereto. The solution was stirred for 30 min and then dried at 100 °C until the gel formed and condensed, and then at a temperature of 250 °C until a powder was formed.

2. Synthesis of BiFeO₃ powder used nitrate salts of Fe and Bi, water, HNO₃, and citric acid as a solvent. The basic compounds were dissolved in HNO₃ acid, which was then heated on a hot plate at 80–90 °C to form a gel (about 4–5 h). The resulting gel was then heated in an oven at a temperature of 180 °C for 2 h. The aim is to evaporate the water.

The annealing temperature for both powders was 550 °C (during 10 h), 600 °C (during 3 h), 700 °C (during 3 h), 800 °C (during 3 h (see Fig. 1).

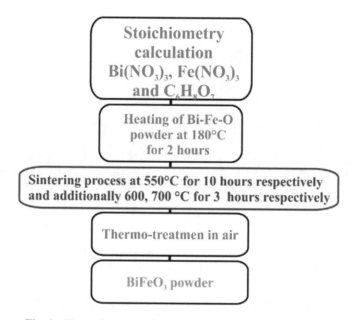

Fig. 1. The main stages of sol-gel synthesis of BiFeO₃ powders

To study the particle size distribution the suspension based on BiFeO₃ powder was prepared. Isopropanol was used as the solvent. The resulting suspension was applied to the surface of single-crystal silicon for carrying out AFM investigation.

3 Results and Discussion

The results of the investigation of the surface of the synthesized powder are shown in Fig. 2 (AFM image) and XRD in Fig. 3. The highest content of the required rhombohedral phase is observed for powder obtained by the first synthesis route annealed for 10 h at a temperature of 550 °C.

The resulting phase is stable. Annealing at the higher temperatures leads to the removal of bismuth from the crystal lattice. Further annealing of the formed material at higher temperatures does not lead to an increase in the content of the required phase.

As can be seen from the XRD data, the BFO reaction product was not monophasic (rhombohedral phase). The determining factor is associated with the peculiarities of the sol-gel synthesis technique. The increasing of the synthesis temperature leads to the decrease in the content of the perovskite phase due to the weak bond of bismuth ions in the crystalline cell (Table 1).

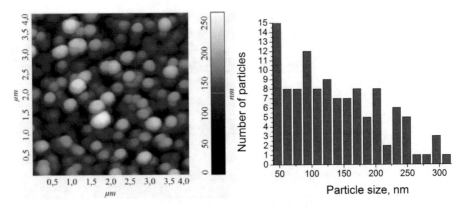

Fig. 2. AFM-image of $BiFeO_3$ powder obtained by sol-gel method (annealing temperature 550 °C)—left image. Particle distribution graph is denoted on the right image

In conclusion, the mixtures of different BFO-related phases with the perovskite phase with a content R-phase up to about 75% were synthesized using sol-gel method.

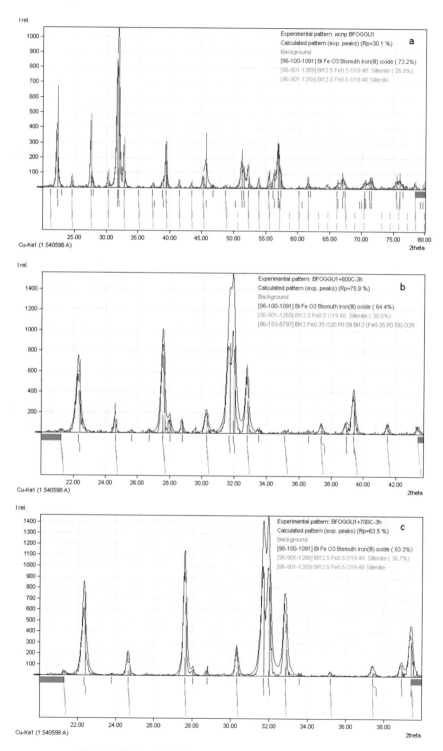

Fig. 3. XRD of BiFeO₃ powder obtained by sol-gel method

Table 1. Phase content of $BiFeO_3$ powder obtained by sol-gel method.

Powder	Temperature and processing time, °C	Phase content $BiFeO_3$, %
BFO	550–10 h	74
	700–3 h	64
	800–3 h	63

Acknowledgements. This project has received funding from the European Union's Horizon 2020 research and innovation programme under the Marie Sklodowska-Curie, grant agreement No 778070 - TransFerr - H2020 - MSCA - RISE-2017.

References

1. Suastiyanti, D., Wijaya, M.: Synthesis of $BiFeO_3$ nanoparticle and single phase by sol-gel process for multiferroic material. ARPN J. Eng. Appl. Sciences **11**(2), 901–905 (2016)
2. Sidsky, V.V., Semchenko, A.V., Khakhomov, S.A., Morozovska, A.N., Morozovsky, N. V.,. Kolos, V.V., Turtsevich, A.S., Pyatlitski, A.N., Pleskachevsky, Yu, M.S., Shil'ko, V., Petrokovets,E.M.: Ferroelectric Properties of Nanostructured SBTN Sol-Gel Layers. Recent Global Research and Education: Technological Challenges. In: Ryszard Jablonski, Roman Szewczyk (eds.) Advances in Intelligent Systems and Computing, vol. 519, pp. 103–108. Springer, Cham (2017)
3. Khakhomov, S.A., Semchenko, A.V., Sidsky, V.V., Gaishun, V.E., Luca, D., Kolos, V.V., Solodukha, V.A., Pyatlitski, A.N., Kovalchuk, N.S.: Nanostructure and Ferroelectric Properties of Sol-Gel SBTN-Films for Electronic Devices. Recent Advances in Technology Research and Education. INTER-ACADEMIA 2017, vol. 660, pp. 144–150. Springer, Cham (2018)

Influence of Deposition Parameters on the Structure of TiO$_2$ Thin Films Prepared by Reactive Magnetron Sputtering Technique

Vytautas Kavaliunas$^{(\boxtimes)}$, Audrone Sestakauskaite, Mantas Sriubas, and Giedrius Laukaitis

Department of Physics, Faculty of Mathematics and Natural Sciences, Kaunas University of Technology, Studentu g. 50, 51368 Kaunas, Lithuania
Vytautas.kavaliunas@ktu.edu

Abstract. TiO$_2$ is well known for its photocatalytic properties and wide range of applications. However, the efficiency of amorphous TiO$_2$ as photocatalyst is low and deposition of crystal TiO$_2$ phases is strict to deposition parameters. The TiO$_2$ phase dependence on temperature, total pressure (p_{tot}) and oxygen partial pressure and total pressure (p_{O2}/p_{tot} [%]) ratio and how this affect growth rate has been studied in this work, TiO$_2$ thin films were deposited via magnetron sputtering technique using different deposition parameters in order to get TiO$_2$ either pure anatase or rutile phase. Crystallographic structure and morphology of deposited thin films were analyzed by XRD and SEM/EDS. TiO$_2$ phase strongly depends on substrate temperature during the deposition of thin films, total pressure and p_{O2}/p_{tot} ratio. Analysis shows that TiO$_2$ anatase phase depends more on substrate temperature than p_{O2}/p_{tot}, while TiO$_2$ rutile phase in reverse, noting that it has better stability at high temperatures compared to anatase.

Keywords: TiO$_2$ · TiO$_2$ characterization · TiO$_2$ phase dependence

1 Introduction

TiO$_2$ stands as one of the most studied material due to its wide range of applications: air/water purification, hydrogen production, self-cleaning coatings, etc. In recent years, TiO$_2$ has been widely studied for its photocatalytical properties and applications for photocatalysis reactions [1–5]. It is well known that crystalline structure of TiO$_2$ achieves better photocatalytic properties than amorphous TiO$_2$ because of regular crystalline structure and lower rates of recombination processes [6, 7]. The phase of TiO$_2$ is an important factor for photocatalysis reaction. Rutile, anatase phases have been investigated for years while amorphous TiO$_2$ not as much [8, 9].

Therefore, crystalline TiO$_2$ thin films have been formed using various techniques and parameters to enhance photocatalytic properties [10–15]. Chemical vapor deposition (CVD), physical vapor deposition (PVD) and Sol-gel dip coating are most common formation methods. The latter is widely used, and thin films structure depends on factors like particle size, substrate withdrawal speed, relative rate of condensation, evaporation, etc. [16]. This technique is less expensive compared to CVD and PVD.

© Springer Nature Switzerland AG 2019
G. Laukaitis (Ed.): INTER-ACADEMIA 2018, LNNS 53, pp. 49–57, 2019.
https://doi.org/10.1007/978-3-319-99834-3_7

Despite that, it has one common issue – purity of thin films. In this case, PVD is more acceptable technique because deposition of thin films is carried out in high vacuum, i.e. the quality of thin films is much higher [17, 18]. For example, reactive magnetron sputtering seems the most promising, because of its versatility, possibility to deposit coatings on the surface with an area of several square meters. Moreover, synthesized coatings exhibit a high adhesion to a substrate. Nevertheless, TiO_2 thin films deposited by magnetron sputtering usually have mixed rutile and anatase phase. It has been many attempts to get pure TiO_2 by changing deposition parameters, i.e. total gas pressure, gas flow ratio, and substrate temperature. To get anatase or rutile phase, TiO_2 samples are usually annealed at > 300 °C temperature or deposited under heating. Amorphous TiO_2 begins to form anatase phase at 300 °C and at ~ 800 °C phase changes to rutile phase [19]. However, deposition of crystalline TiO_2 thin films on heat sensitive substrates requires to lower the temperature of sputtering process. It can be achieved by changing total pressure, and oxygen partial pressure and total pressure (p_{O2}/p_{tot}) ratio. Total gas pressure has influence on the kinetic energy of sputtered atoms. An increase in the total pressure increases the density of gas particles in the chamber and decreases the cathode potential, which influences the probability of collisions and the acceleration of particles and consequently the particle energy [20]. Therefore, rutile phase forms at low total pressure and anatase phase forms at high total pressure [20, 21]. Gas flow ratio also has an influence on the phase formation in TiO_2 thin films. O^- ion concentration increases in vacuum chamber with increasing p_{O2}/p_{tot} ratio [22]. Negative oxygen ions are accelerated by the cathode sheath and move towards the substrate. The bombardment effect occurs which is a possible reason for the production of rutile phase [23]. Most of investigations analyze the influence p_{O2}/p_{tot} > 20%. However, the deposition rate suffers strong decrease with increasing oxygen partial pressure. Therefore, optimal p_{O2}/p_{tot} ratio should be found to get the highest possible deposition rate and the purest anatase or rutile phase.

In the present work, TiO_2 thin films have been deposited under different conditions to study phase dependence on deposition parameters, i.e. p_{O2}/p_{tot} ratio (10–20%) and substrate temperature during the deposition (25–300 °C).

2   Experimental

2.1   Sample Preparation and Characterization

TiO_2 layers were deposited using PVD 75 vacuum system manufactured by Kurt J. Lesker company and consisted of two stage rotary vane fore vacuum pump and turbomolecular vacuum pump achieving pressure to 6.6×10^{-5} Pa. Argon and oxygen gases (with 99.999% purity) were used in deposition during plasma activation. Two titanium (Ti) target (99.995% purity) with diameter of 50.8 mm and thickness of 6.35 mm were used in process. DC (direct current) power source (200 W) was used for Ti targets. TiO_2 layers were deposited on Si substrate in chamber which was pumped to 6.6×10^{-3} Pa before every deposition process. The chamber than was filled with Ar/O ratio of 90/10, 86/14, 84/16 and 80/20, respectively. Deposition process were done when pressure in chamber stabilizes to 0.79 Pa work pressure (the pressure was oxygen concentration dependent) and was set in room temperature (25 °C), 100, 200, and 300 °C. Substrate and targets were separated with shutter of the magnetron between them while etching process were set to clean the target surface from additional oxygen. Deposition time varies from 2 to 6 h (depends on argon and oxygen ratio and temperature) to achieve 200–300 nm thickness of samples.

The phase composition of thin films was determined with an X-ray diffraction measurement at grazing incidence (XRDGI) using a D8 Discover diffractometer (Bruker) with Cu K$_\alpha$ (λ = 1.54 Å) X-ray source. X-ray generator voltage and current was 40.0 kV and 40 mA, respectively. The XRDGI scans were performed in the range of 4°–134.0°. The XRDGI scans were performed at incidence angle of 1.0°. The surface area was analyzed with scanning electron microscope (SEM) and elemental analysis was set with energy dispersive X-ray spectroscopy (EDS).

3   Results and Discussion

TiO$_2$ thin films deposition time was adjusted to obtain the thickness of thin films between 200–300 nm. The growth rate of thin films depends on oxygen gas concentration (Fig. 1). The decrease of deposition rate occurs due to oxidation of Ti target. Larger area of oxidized Ti target at higher oxygen flow rate and much lower sputtering yield of TiO$_x$ decrease deposition rate [24].

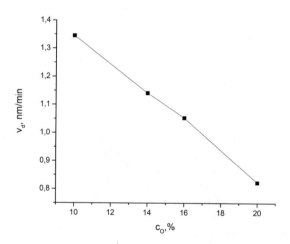

Fig. 1. TiO$_2$ thin films growth rate dependence on oxygen gas

In accordance to the penetration depth of UV light, which is approximately 120–180 nm (depending on the dielectric constant of thin films), deposition time was set accordingly to deposition rate to achieve thickness no less than 200 nm. Configuration of optimal oxygen concentration for pure anatase and rutile phases, low temperature for more economical deposition and low total pressure for high deposition rates must be analyzed.

XRD measurements were performed to study the structural properties of TiO_2 thin films and the influence of the deposition parameters on the formed thin films. Peaks of (101); (004); (200) crystallographic orientations at angles 25.54°; 37.77°; 48.02° respectively confirm the TiO_2 anatase structure [25] and peaks of (110); (101) crystallographic orientation at angles 27.52°; 35.83° respectively confirm rutile structure (Fig. 2) [26]. Other peaks show monoclinic non-stoichiometric TiO_x structure with different orientation. Since TiO_2 anatase peak at 25.54° and rutile at 27.52° are the most important to analyze crystallinity of thin films, further analysis was done in the range of 20°–35°.

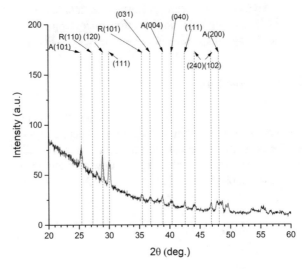

Fig. 2. XRD patterns of TiO_2 thin films deposited under 300 °C and p_{O2}/p_{tot} ratio of 10%

XRD analysis revealed that TiO_2 thin films deposited on room temperature substrates have mixture of rutile and TiO_x phases (Fig. 3a). The intensity of rutile peak (27.52°) increases with increasing p_{O2}/p_{tot} ratio. Finally, peaks of TiO_x vanish and pure rutile phase forms when p_{O2}/p_{tot} ratio is 20%. In the sputtering process, high energy Ti atoms and negative oxygen ions bombard the film surface and affect the structure. Bombardment is dependent of p_{O2}/p_{tot} ratio because the kinetic energy of sputtered particles decreases while increasing the p_{O2}/p_{tot} ratio [27, 28]. It is known that rutile TiO_2 has larger density than anatase TiO_2 (R—4.250 g/cm^3 and A—3.894 g/cm^3) and bombardment of high energy particles induce the formation of rutile phase. Despite that, Kotake et al. suggests that with decreased kinetic energy of Ti atoms, migration energy increases which affect the growing surface causing the densification of TiO_2 to form rutile phase [29]. Moreover, Safeen et al. studies shows that the bombardment of Ar^+ and Ti^+ ions generates oxygen vacancies which leads to the crystallization of rutile phase at low temperature [22]. Many studies show that that rutile phase forms either at high temperatures or at low total pressure (~ 0.2 Pa) and high p_{O2}/p_{tot} ratio (from 30%) [20]. If TiO_2 thin films are deposited at room temperature, the formation of anatase or rutile phases are possible only after annealing TiO$_2$ at 450 °C or at 600–800 °C temperatures respectively [19]. TiO$_2$ thin films deposited on 300 °C substrates have mixture of anatase, rutile and TiO$_x$ phases (Fig. 3b). TiO$_2$ anatase peak (25.54°) has higher intensity than rutile peak (27.52°). Moreover, intensity of both peaks increases with increasing p_{O2}/p_{tot} ratio. However, TiO$_x$ peaks do not disappear when p_{O2}/p_{tot} ratio is 20% in comparison to thin films deposited at room temperature.

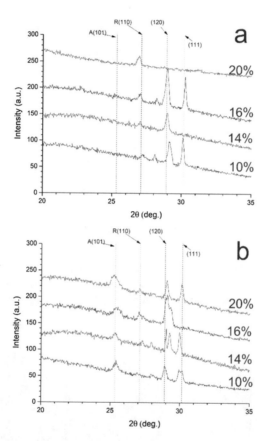

Fig. 3. XRD patterns of deposited TiO$_2$ thin films with different p_{O2}/p_{tot} ratio and deposited without additional heating (a) and under 300 °C heating of substrate (b)

It means that deeper analysis is required considering the substrate temperature (Fig. 4). As mentioned earlier, pure rutile forms at room temperature (p_{O2}/p_{tot} ratio – 20%). The intensity of rutile (110) peak decrease and the peak of TiO$_x$ appears using 100 °C substrate temperature. The second peak of TiO$_x$ structure appears and the intensity of the rutile peak decrease further at 200 °C. Finally, at 300 °C substrate temperature, anatase peak appears. It is known that the grain size increases with temperature and the increased grain size caused the formation of anatase phase [30]. Pure TiO$_2$ anatase forms at 450 °C. Thus, suggests that TiO2 rutile phase is more

p_{O2}/p_{tot} ratio dependent than on substrate temperature (noting that rutile is more stable at high temperatures than anatase) and TiO_2 anatase phase is more dependant on substrate temperature than on p_{O2}/p_{tot}.

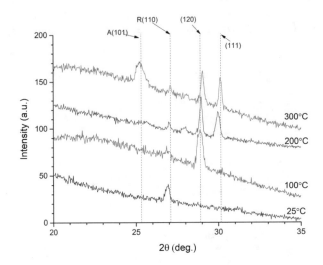

Fig. 4. XRD patterns of 20% p_{O2}/p_t ratio at different substrate temperatures

SEM images show that TiO_x thin films are homogeneous, without cracks and pores (Fig. 5). The grains are very small. Their size increases with increasing substrate temperature during deposition. Using higher substrates temperature, adatoms have

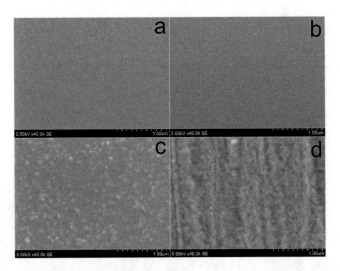

Fig. 5. SEM images of TiO_2 thin films deposited at various substrate temperatures, i.e. 25 °C (a), 100 °C (b), 200 °C (c) and 300 °C (d)

longer diffusion path and time. Therefore, larger grains are formed. Still, small grains and low substrate temperature clearly indicates that thin films grow in Zone I according Thornton structure zone model [31].

4 Conclusion

In this study, TiO_2 films under various deposition parameters, using magnetron sputtering system, were deposited on Si substrate. During the sputtering deposition temperature, total pressure and p_{O2}/p_{tot} ratio obviously affects the crystallographic structure of thin films. TiO_2 anatase phase depends more on substrate temperature than on p_{O2}/p_{tot} ratio while TiO_2 rutile depends more on p_{O2}/p_{tot} ratio than on substrate temperature. This dependence of rutile structure was considered to be caused by the bombardment effect of high-energy particles or migration of Ti atoms on the surface. Anatase dependence on substrate temperature is caused by increased grain size with increase of temperature. Also, need to mention that anatase transforms to rutile at high temperatures (600–800 °C). Thus, could be explained by density of structures knowing that rutile is denser than anatase. Nevertheless, further investigation of thin films deposition needs to be done to get optimal configuration for magnetron sputtering technique in order to get TiO_2 anatase or rutile structure.

References

1. Binas, V., Venieri, D., Kotzias, D.: Modified TiO_2 based photocatalysts for improved air and health quality. J. Mater. **3**, 3–16 (2017)
2. Abdullah, H., Khan, M.M.R., Ong, H.R., Yaakob, Z.: Modified TiO_2 photocatalyst for CO_2 photocatalytic reduction: an overview. J. CO2 Util. **22**, 15–32 (2017)
3. Daghrir, R., Drogui, P., Robert, D.: Modified TiO_2 for environmental photocatalytic applications: a review. Ind. Eng. Chem. Res. **52**, 3581–3599 (2013)
4. Kumar, S.G., Devi, L.G.: Review on modified TiO_2 photocatalysis under UV/visible light: selected results and related mechanisms on interfacial charge carrier transfer dynamics. J. Phys. Chem. A **115**, 13211–13241 (2011)
5. Kaneco, S., Shimizu, Y., Ohta, K., Mizuno, T.: Photocatalytic reduction of high pressure carbon dioxide using TiO_2 powders with a positive hole scavenger. J. Photochem. Photobiol. A Chem. **115**, 223–226 (1998)
6. Damm, C., Herrmann, R., Israel, G., Müller, F.W.: Acrylate photopolymerization on heterostructured TiO_2 photocatalysts. Dye. Pigment. **74**, 335–342 (2007)
7. Manzanares, M., et al.: Engineering the TiO_2 outermost layers using magnesium for carbon dioxide photoreduction. Appl. Catal. B Environ. **150–151**, 57–62 (2014)
8. Rui, Z., Wu, S., Peng, C., Ji, H.: Comparison of TiO_2 with anatase and rutile crystalline phases for methane combustion. Chem. Eng. J. **243**, 254–264 (2014)
9. Bakardjieva, S., Subrt, J., Stengl, V., Dianez, M.J., Sayagues, M.J.: Photoactivity of anatase–rutile TiO_2 nanocrystalline mixtures obtained by heat treatment of homogeneously precipitated anatase. Appl. Catal. B Environ. **58**, 193–202 (2005)
10. Dette, C., et al.: TiO_2 anatase with a bandgap in the visible region. Nano Lett. **14**, 6533–6538 (2014)

11. Wan, L., Li, J.F., Feng, J.Y., Sun, W., Mao, Z.Q.: Anatase TiO_2 films with 2.2 eV band gap prepared by micro-arc oxidation. Mater. Sci. Eng. B **139**, 216–220 (2007)
12. Shang, C., Zhao, W.-N., Liu, Z.-P.: Searching for new TiO_2 crystal phases with better photoactivity. J. Phys. Condens. Matter **27**, 134203 (2015)
13. Reyes-Coronado, D., et al.: Phase-pure TiO_2 nanoparticles: anatase, brookite and rutile. IOP Publ. Nanotechnol. Nanotechnol. **19**, 145605–145610 (2008)
14. De Angelis, F., Di Valentin, C., Fantacci, S., Vittadini, A., Selloni, A.: Theoretical studies on anatase and less common TiO_2 phases: bulk, surfaces, and nanomaterials. Chem. Rev. **114**, 9708–9753 (2014)
15. Chen, X., Mao, S.S.: Titanium dioxide nanomaterials: synthesis, properties, modifications, and applications. Chem. Rev. **107**, 2891–2959 (2007)
16. Brinker, C.J., Frye, G.C., Hurd, A.J., Ashley, C.S.: Fundamentals of sol-gel dip coating. Thin Solid Films **201**, 97–108 (1991)
17. Li, W., Ni, C., Lin, H., Huang, C.P., Shah, S.I.: Size dependence of thermal stability of TiO_2 nanoparticles. https://doi.org/10.1063/1.1807520
18. Cernuto, G., Masciocchi, N., Cervellino, A., Colonna, G.M., Guagliardi, A.: Size and shape dependence of the photocatalytic activity of TiO_2 nanocrystals: a total scattering debye function study. J. Am. Chem. Soc. **133**, 3114–3119 (2011)
19. Taherniya, A., Raoufi, D.: The annealing temperature dependence of anatase TiO_2 thin films prepared by the electron-beam evaporation method. Semicond. Sci. Technol. **31**, 125012 (2016)
20. Zeman, P., Takabayashi, S.: Effect of total and oxygen partial pressures on structure of photocatalytic TiO_2 films sputtered on unheated substrate. Surf. Coat. Technol. **153**, 93–99 (2002)
21. Buranawong, A., Witit-anun, N., Chaiyakun, S.: Total pressure and annealing temperature effects on structure and photo-induce hydrophilicity of reactive DC sputtered TiO_2 thin films. Eng. J. **16**, 79–89 (2012)
22. Safeen, K., Micheli, V., Bartali, R., Gottardi, G., Laidani, N.: Low temperature growth study of nano-crystalline TiO_2 thin films deposited by RF sputtering. J. Phys. D Appl. Phys. **48**, 295201 (2015)
23. Jia, J., Yamamoto, H., Okajima, T., Shigesato, Y.: On the crystal structural control of sputtered TiO_2 thin films. Nanoscale Res. Lett. **11**, 324 (2016)
24. Baroch, P., Musil, J., Vlcek, J., Nam, K.H., Han, J.G.: Reactive magnetron sputtering of TiO_x films. Surf. Coat. Technol. **193**, 107–111 (2005)
25. Theivasanthi, T., Alagar, M.: Titanium dioxide (TiO_2) Nanoparticles -XRD Analyses—An Insight
26. Bokhimi, X., Morales, A., Pedraza, F.: Crystallography and crystallite morphology of rutile synthesized at low temperature. J. Solid State Chem. **169**, 176–181 (2002)
27. Okimura, K.: Low temperature growth of rutile TiO_2 films in modified rf magnetron sputtering. Surf. Coat. Technol. **135**, 286–290 (2001)
28. Burdett, J.K., Hughbanks, T., Miller, G.J., Richardson, J.W., Smith, J.V.: Structural-electronic relationships in inorganic solids: powder neutron diffraction studies of the rutile and anatase polymorphs of titanium dioxide at 15 and 295 K. J. Am. Chem. Soc. **109**, 3639–3646 (1987)
29. Kotake, H., Jia, J., Nakamura, S., Okajima, T., Shigesato, Y.: Tailoring the crystal structure of TiO_2 thin films from the anatase to rutile phase. J. Vac. Sci. Technol. A Vacuum Surf. Film. **33**, 041505 (2015)

30. Liu, G.X., Shan, F.K., Lee, W.J., Shin, B.C.: Growth temperature dependence of TiO$_2$ thin films prepared by using plasma-enhanced atomic layer deposition method. J. Korean Phys. Soc. **50**, 1827 (2007)
31. Kluth, O., Schope, G., Hupkes, J., Agashe, C., Muller, J.: Modified Thornton model for magnetron sputtered zinc oxide: film structure and etching behaviour. Thin Solid Films **442**, 80–85 (2003)

Gamma-Ray Spectroscopic Performance of Large-Area CdTe-Based Schottky Diodes

Volodymyr A. Gnatyuk[1,2](✉) 📧, Kateryna S. Zelenska[1,3],
Valery M. Sklyarchuk[4], Wisanu Pecharapa[5], and Toru Aoki[1]

[1] Research Institute of Electronics, Shizuoka University,
3-5-1 Johoku, Naka-ku, Hamamatsu 432-8011, Japan
gnatyuk@ua.fm

[2] V.E. Lashkaryov Institute of Semiconductor Physics, National Academy
of Sciences of Ukraine, Prospekt Nauky 41, Kiev 03028, Ukraine

[3] Faculty of Physics, Taras Shevchenko National University of Kyiv,
Prospekt Akademika Glushkova 4, Kiev 03127, Ukraine

[4] Yuriy Fedkovych Chernivtsi National University,
Kotsubynskyi Str. 2, Chernivtsi 58012, Ukraine

[5] College of Nanotechnology, King Mongkut's Institute of Technology
Ladkrabang, 1 Thanon Chalong Krung, Ladkrabang, Bangkok 10520, Thailand

Abstract. Spectroscopic performance of the Ni/CdTe/Au Schottky diode X/γ-ray detectors was examined by measurements of ^{137}Cs isotope spectra at different bias voltages and operation time. Both the Schottky (Ni-CdTe) and near Ohmic (Au-CdTe) contacts were formed on the opposite sides (10×10 mm^2) of high resistivity CdTe(111) crystals after preliminary chemical etching and Ar-ion bombardment with different parameters. The detectors had low reverse dark current density (4–6 nA/cm^2 and 8–12 nA/cm^2 at bias of 500 V and 1000 V, respectively) and showed quite high energy resolution (2–4%@662 keV) at room temperature. The optimal bias voltage ranges for the Ni/CdTe/Au detectors were determined to achieve sufficiently high detection efficiency and energy resolution. These parameters were changing by 30–50% for the detectors biased and subjected to γ-ray radiation during several hours that evidenced quite low degradation. The detector spectroscopic parameters can be recovered by turning off the bias for seconds and turned on it again to continue measurements.

Keywords: CdTe crystals · Schottky diodes · X/γ-ray detectors
Isotope spectra

1 Introduction

Various materials, particularly metals are used for creation of electrical contacts and formation of electrodes in the fabrication of CdTe-based X/γ-ray detectors [1–15]. The choice of a certain metal depends on the purpose to obtain a Schottky (rectifying) or Ohmic contact to p- or n-type of CdTe crystals. It is known, that diode-type detectors allow obtaining much higher energy resolution compared with detectors having both Ohmic contacts despite the lower stability of the diode detector parameters.

© Springer Nature Switzerland AG 2019
G. Laukaitis (Ed.): INTER-ACADEMIA 2018, LNNS 53, pp. 58–65, 2019.
https://doi.org/10.1007/978-3-319-99833-3_8

The need for room temperature X/γ-ray imaging detectors with large area and high energy resolution, in particular for astrophysical and medical applications, has caused searching and developing efficient techniques to fabricate CdTe-based diode-type detectors either as Schottky diodes [1–11] or rectifying structures with an electrical junction (p-i-n or M-p-n structures) [1, 10–15]. In general, In (Al) and Au (Pt) are commonly used for high-resistivity detector grade p-like CdTe crystals to create a blocking and Ohmic contact, respectively [2–8]. However, there are much more materials which are interesting and can be promising for creation of contacts and electrode deposition in CdTe diode-type detector fabrication.

The choice of Ni for creation of a Schottky contact in our study was due to the suitable parameters of the metal and semiconductor. Ni has not been employed as an electrical contact metal in CdTe-based high energy detector fabrication so commonly as In, Al, Au or Pt because of the necessity of specific preparation of the semiconductor crystal surface before the electrode deposition and peculiarities of the Ni deposition procedures [7–10]. However, Ni can be an attractive material for creation of electrical contacts, particularly for large area electrode formation thanks to its properties. Along with appropriate electronic characteristics (high work function ∼4.7 eV, etc.), Ni is chemically low-active, ductile and hard metal. Ni contacts (electrodes) are mechanically strong and not corroded during long time operation and storage in various environment conditions. This allows avoiding addition procedures to protect the electrode which are used in the industrial (Acrorad) fabrication of CdTe Schottky diode detectors with an In contact and include coating an In electrode with Ti in order to prevent corrosion and damage [2–5].

In this study, the γ-ray spectroscopic characteristics of the diode-type detectors with the surface area of 10×10 mm^2, developed on the base of commercial (Acrorad) semi-insulating CdTe crystals by formation of a high barrier Schottky contact at the Ni-CdTe interface and near Ohmic contact at the Au-CdTe interface, using ion etching in both the cases, were measured and discussed.

2 Experimental Details

The Ni/CdTe/Au Schottky diode X/γ-ray detectors were fabricated based on detector-grade Cl-compensated (111) oriented p-like CdTe semiconductor with the resistivity of 4–6×10^9 Ω cm produced by Acrorad Co., Ltd. [5]. Parallelepiped-like CdTe(111) single-crystal wafers with the sizes of $10 \times 10 \times 0.75$ mm^3, polished by the manufacturer, were employed. Both the rectifying (Ni-CdTe) and Ohmic (Au-CdTe) contacts were formed after preliminary chemical etching of the CdTe crystals in a Br-methanol solution and following treatment with an Ar plasma at different regimes (ion energy, beam density, processing duration, etc.) for the B- and A-faces, respectively.

A Ni contact was created on the CdTe(111)B surface by vacuum evaporation using a Mo mask with sizes of approximately 8×8 mm^2. An Au contact was formed on the entire CdTe(111)A crystal surface by chemical deposition from a gold chloride solution. The area of the Ni (rectifying contact) and Au (near Ohmic contact) electrodes was around 0.6 cm^2 and 1 cm^2, respectively. Passivation of the lateral faces of the Ni/CdTe/Au diodes was used to eliminate leakage current.

The obtained Schottky diodes operated when the Ni contact was positively biased with respect to the Au one (reverse mode) and showed quite low reverse room temperature dark currents even at high bias voltages (current density was 4–6 nA/cm^2 and 8–12 nA/cm^2 at 500 V and 1000 V, respectively). The Ni/CdTe/Au Schottky diode detectors were tested by spectroscopic measurements using a ^{137}Cs isotope as a radiation source and compact portable spectrometer ANS-MNT004-GTK with the form-factor of a USB stick, fabricated by ANSeeN Inc. [16]. The Ni/CdTe/Au detectors were exposed to ^{137}Cs γ-ray radiation from the Ni electrode side and the spectra were taken at applied bias voltage V = 100–1000 V with the measurement time of 5 min or 30 min in dark at room temperature. The distance between the radiation source and detector was 30 cm and it was not changed during the spectral measurements.

3   Results and Discussion

Diode-type X/γ-ray detectors generally operate in reverse bias mode, therefore it was important that the fabricated Ni/CdTe/Au structures had low reverse dark current even at high bias voltages. This was accomplished by the developed electrical contact formation technology and allowed increasing bias voltage to extend the depletion region up to whole thickness of the CdTe crystal and thus, better or even full collection of photogenerated charge carriers was achieved while avoiding electrical breakdown.

The energy spectra of a ^{137}Cs isotope taken by the Ni/CdTe/Au Schottky diode detector at applied bias voltage V = 1000 V during 5 min and 30 min are shown in Fig. 1 (curves 1 and 2, respectively). The prominent 662 keV line typical for the ^{137}Cs spectrum is clearly observed. The symmetric shape of this line indicated the full charge collection at the chosen bias voltage. The broad shoulder at the low-energy side from the 662 keV line was attributed to Compton scattering of γ-rays [4, 5].

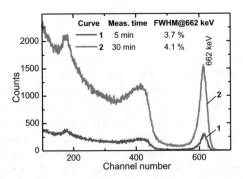

Fig. 1. Spectra of a ^{137}Cs isotope taken with the Ni/CdTe/Au Schottky diode detector at applied bias voltage of 1000 V and with measurement time of 5 min (curve 1) and 30 min (curve 2)

The full width at half-maximum (FWHM) of the 662 keV photopeak, characterizing the detector energy resolution, is 3.7% in the case of 5 min measurement (Fig. 1, curve 1). Attention is drawn to a significant increase in the photopeak intensity (number of counts) when the spectrum was measured during 30 min (Fig. 1, curve 2). This directly proportional increase in the number of counts over time evidences high performance of the Ni/CdTe/Au Schottky diode detector.

A slight broadening of the 662 keV line (Fig. 1, curve 2, FWHM = 4.1%), compared with that in the spectrum taken during 5 min (Fig. 1, curve 1, FWHM = 3.7%), is associated with the accumulation of charge carriers in the CdTe crystal surface layer at the Schottky contact during longer measurement time that leads to a change in the electric field distribution in the space charge region [6]. This phenomenon, so-called "polarization effect" is an inevitable effect for both kinds of CdTe diode-type X/γ-ray detectors: with a Schottky contact, formed using various metals [3, 6, 9, 11], and barrier structures with a *p-n* junction [11, 12]. However, the detection efficiency of the Ni/CdTe/Au detector did not decrease during 30 min operation.

The long-time stability of the spectral characteristics is a key requirement for spectroscopic application of room temperature CdTe-based X/γ-ray detectors, in particular it is very important for diode-structured devices [1–15]. In order to study the capabilities of the fabricated Ni/CdTe/Au diode to operate long time without significant deterioration of its functional parameters, the spectra of a ^{137}Cs isotope were measured just after the detector was biased and after operating during different times when the detector was kept under applied bias voltage and γ-ray irradiation. Figure 2 shows the ^{137}Cs isotope spectra, all measured at V = 1000 V during 5 min, at the beginning of γ-ray detection (curve 1) then, after 60 min (curve 2) and 120 min (curve 3) operation.

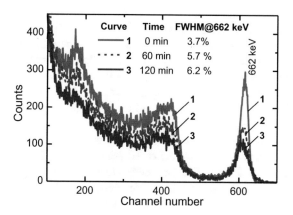

Fig. 2. Spectra of a ^{137}Cs isotope taken with the Ni/CdTe/Au Schottky diode detector at applying bias voltage of 1000 V (curve 1) and after operation (under bias and γ-ray irradiation) during 60 min (curve 2) and 120 min (curve 3). Spectrum measurement time was 5 min

As seen from the evolution of the energy spectra of the Ni/CdTe/Au detector after the bias voltage applied, the 662 keV photopeak height decreased twice in 60 min and energy resolution worsen, i.e. FWHM of the 662 keV line changed from 3.7% to 5.7% (Fig. 2, curve 2).

These parameters (number of counts and FWHM) continued to change with time but to a lesser extend (Fig. 2, curve 3). The energy position of the maximum of the spectra was slightly shifted from the peak channel position of the 662 keV line from a 137Cs isotope (Fig. 2, curves 2 and 3). The time-dependent changes in the spectral characteristics evidenced that the Ni/CdTe/Au Schottky diode detector still suffered from the polarization due to prolonged electrical bias and exposure to γ-ray radiation. However, a tendency toward saturation in changes of the detection efficiency and energy resolution in the 137Cs isotope spectra with time was observed and the 662 keV line was clearly detected even after 12–15 h of detector operation.

For spectroscopic and imaging applications, severe degradation of isotope spectra should be avoided, so the detector is required to be recovered from the polarization. It was possible by interrupting the bias voltage. Electrical bias, applied to the detector, was turned off for a few seconds and turned on again to measure the spectrum at the same parameters ($V = 1000$ V and measurement time of 5 min). The obtained spectrum was almost the same as the initial one (Fig. 2, curve 1). Thus, for long-time stable operation of the developed Ni/CdTe/Au detectors, short-time interrupting the bias voltage can be applied to provide recovering the spectral characteristics.

The recovery of the electrical and spectral parameters was usually observed for CdTe-based diode-type detectors when the bias was switched off for a few seconds [9] or after the bias voltage was re-applied to a detector [3, 11]. Further investigations are necessary to study the time-dependence features of the spectroscopic characteristics of the fabricated Ni/CdTe/Au Schottky diode detectors for increasing degradation time and thus their stability.

The ^{137}Cs isotope spectra strongly depended on the applied bias voltage that is illustrated by the data presented in Fig. 3. The number of counts, position and FWHM of the 662 keV photopeak are different at different bias voltages (Fig. 3, curves 1–3). The broad shoulder and tail, extending toward the low-energy side from the 662 keV peak, were due to incomplete of charge collection (Fig. 3, curve 1). A part of photo-generated carriers was trapped in the detector bulk at low bias ($V = 100$ V) [2–4].

With increasing V from the lowest applied value (100 V) to a few hundred volts, the number of counts and energy resolution increased significantly, however at $V > 400$–500 V, a tendency toward saturation of these characteristics was observed. At higher voltages $V > 800$–900 V, the detection efficiency began to slightly decrease and FWHM of the 662 keV photopeak increased (Fig. 3, curve 3). Such a trend of the voltage dependence of the FWHM was determined by the competition of two processes. When bias voltage increased, the charge collection efficiency also increased and the photopeak FWHM decreased. At the same time with rising V, the reverse current of the Schottky diode increased that led to an increase in electrical noise and thus to an increase in the FWHM of the line in the isotope emission spectrum [9].

Studying the electric field strength dependence of the ^{137}Cs isotope energy spectrum, it was established that the applied bias $V = 700$ V was the optimal operation voltage for the investigated Ni/CdTe/Au detector because the peak channel position of the 662 keV line reached its maximum and the FWHM of this line achieved the minimum value (3.3%) as well as the number of counts became highest one demonstrating the best spectroscopic performance (Fig. 3, curve 2).

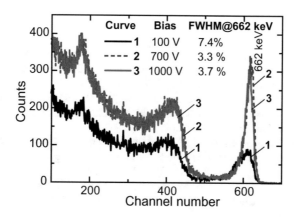

Fig. 3. Spectra of a ^{137}Cs isotope taken with the Ni/CdTe/Au Schottky diode detector at applied bias voltages V (V): 100 (curve 1), 700 (curve 2) and 1000 (curve 3). Spectrum measurement time was 5 min

There was a certain optimal bias voltage range for each Ni/CdTe/Au Schottky diode detector to obtain the best detection efficiency (highest number of counts), high energy resolution (lowest value of the FWHM of the 662 keV line) and true peak channel position (662 keV). The developed Ni/CdTe/Au detectors showed quite good time stability of the spectroscopic characteristics e.g., the number of counts and energy resolution decreased and stabilized in a few hours and then unchanged during long operation time under γ-ray irradiation at room temperature.

4   Conclusion

The CdTe-based X/γ-ray detectors with quite large (10 × 10 mm^2) area were developed as diode structures with the Schottky and near Ohmic contacts both formed using the Ar-ion bombarding technique. The spectroscopic study of the Ni/CdTe/Au Schottky diode detectors revealed the advantageous features. Despite the relatively thin (0.75 mm) CdTe crystals thus, low absorption of high energy radiation (662 keV) hence, low probability of γ-photon capture and generation of electron-hole pairs, the detectors demonstrated quite high detection efficiency that was due to sharp rectification of the Ni/CdTe Schottky contact allowing to apply high bias voltage (up to 1000 V) and thus, to provide full collection of photogenerated charge carriers. Quite satisfactory energy resolution and relatively high degradation resistance under γ-ray irradiation make the detectors promising for spectroscopic and imaging applications.

Acknowledgements. This research was supported by the following international programs and collaborative projects: the NATO Science for Peace and Security Programme (Project SENERA, SfP-984705); Academic Melting Pot Program at King Mongkut's Institute of Technology Ladkrabang (KMITL), Bangkok, Thailand; Bilateral Cooperative Program of 2018 Cooperative Research at Research Center of Biomedical Engineering, Japan (Grant No. 2035).

References

1. Sordo, S.D., Abbene, L., Caroli, E., Mancini, A.M., Zappettini, A., Ubertini, P.: Progress in the development of CdTe and CdZnTe semiconductor radiation detectors for astrophysical and medical applications. Sensors **9**(5), 3491–3526 (2009)
2. Tanaka, T., Kobayashi, Y., Mitani, T., Nakazawa, K., Oonuki, K., Sato, G., Takahashi, T., Watanabe, S.: Recent achievements of the high resolution Schottky CdTe diode for γ-ray detectors. N. Astron. Rev. **48**(1–4), 309–313 (2004)
3. Takahashi, T., Paul, B., Hirose, K., Matsumoto, S., Ohno, R., Ozaki, T., Mori, K., Tomita, Y.: High-resolution Schottky CdTe diode for hard X-ray and gamma-ray astronomy. Nucl. Instrum. Methods Phys. Res. A **436**(1–2), 111–119 (1999)
4. Sato, G., Takahashi, T., Sugiho, M., Kouda, M., Mitani, T., Nakazawa, K., Okada, Y., Watanabe, S.: Characterization of CdTe/CdZnTe detectors. IEEE Trans. Nucl. Sci. **49**(3), 1258–1263 (2002)
5. Shiraki, H., Funaki, M., Ando, Y., Kominami, S., Amemiya, K., Ohno, R.: Improvement of the productivity in the THM growth of CdTe single crystal as nuclear radiation detector. IEEE Trans. Nucl. Sci. **57**(1), 395–399 (2010)
6. Farella, I., Montagna, G., Mancini, A.M., Cola, A.: Study on instability phenomena in CdTe diode-like detectors. IEEE Trans. Nucl. Sci. **56**(4), 1736–1742 (2009)
7. Kosyachenko, L.A., Maslyanchuk, O.L., Sklyarchuk, V.M., Grushko, E.V., Gnatyuk, V.A., Aoki, T., Hatanaka, Y.: Electrical characteristics of Schottky diodes based on semi-insulating CdTe single crystals. J. Appl. Phys. **101**(1), 013704-1-6 (2007)
8. Sklyarchuk, V.M., Gnatyuk, V.A., Pecharapa, W.: Low leakage current Ni/CdZnTe/In diodes for X/γ-ray detectors. Nucl. Instrum. Methods Phys. Res. A **879**, 101–105 (2018)
9. Kosyachenko, L.A., Aoki, T., Lambropoulos, C.P., Gnatyuk, V.A., Sklyarchuk, V.M., Maslyanchuk, O.L., Grushko, E.V., Sklyarchuk, O.F., Koike, A.: High energy resolution CdTe Schottky diode γ-ray detectors. IEEE Trans. Nucl. Sci. **60**(4), 2845–2852 (2013)
10. Lambropoulos, C.P., Aoki, T., Crocco, J., Dieguez, E., Disch, C., Fauler, A., Fiederle, M., Hatzistratis, D.S., Gnatyuk, V.A., Karafasoulis, K., Kosyachenko, L.A., Levytskyi, S.N., Loukas, D., Maslyanchuk, O.L., Medvids, A., Orphanoudakis, T., Papadakis, I., Papadimitriou, A., Potiriadis, C., Schulman, T., Sklyarchuk, V.M., Spartiotis, K., Theodoratos, G., Vlasenko, O.I., Zachariadou, K., Zervakis, M.: The COCAE detector: an instrument for localization - identification of radioactive sources. IEEE Trans. Nucl. Sci. **58**(5), 2363–2370 (2011)
11. Gnatyuk, V.A., Vlasenko, O.I., Aoki, T., Koike, A.: Characteristics and stability of diode type CdTe-based X-ray and gamma-ray detectors. In: 2014 IEEE Nuclear Science Symposium and Medical Imaging Conference Record (NSS/MIC), Seattle, WA, USA, pp. 1–4. IEEE (2016)
12. Gnatyuk, V.A., Aoki, T., Vlasenko, O.I., Levytskyi, S.N., Hatanaka, Y., Lambropoulos, C.P.: Features of characteristics and stability of CdTe nuclear radiation detectors fabricated by laser doping technique. In: Proceedings of SPIE 7079, San Diego, CA, USA, pp. 70790G-1–70790G-9. SPIE (2008)
13. Zelenska, K.S., Gnatyuk, D.V., Aoki, T.: Modification of the CdTe-In interface by irradiation with nanosecond laser pulses through the CdTe crystal. J. Laser Micro/Nanoeng. **10**(3), 298–303 (2015)

14. Gnatyuk, V.A., Levytskyi, S.N., Vlasenko, O.I., Aoki, T.: Formation of doped nano-layers in CdTe semiconductor crystals by laser irradiation with nanosecond pulses. Thai J. Nanosci. Nanotechnol. **1**(2), 7–16 (2016)

15. Aoki, T., Gnatyuk, V.A., Kosyachenko, L.A., Maslyanchuk, O.L., Grushko, E.V.: Transport properties of CdTe X/γ-ray detectors with p-n junction. IEEE Trans. Nucl. Sci. **58**(1), 354–358 (2011)

16. ANSeeN Inc. http://anseen.com. Accessed 25 June 2018

Ellipsometric Diagnostic of Anisotropy Properties of Surface Layer of Silicon After Laser Treatment

Toru Aoki[1], Dmytro Gnatyuk[2], Ludmila Melnichenko[2],
Leonid Poperenko[2], and Iryna Yurgelevych[2(✉)]

[1] Research Institute of Electronics, Shizuoka University, 3-5-1 Johoku,
Hamamatsu 432-8011, Japan
[2] Faculty of Physics, Taras Shevchenko National University of Kyiv,
Academician Glushkov Avenue 2, Building 1, Kiev 03680, Ukraine
vladira_19@ukr.net

Abstract. Optical properties of non-treated silicon plates and Si plates modified by femtosecond laser irradiation have been investigated by ellipsometry. The samples of the nanostructured silicon as isolated cells were formed on the single-crystal silicon wafers by a method of the laser ablation. Laser beam scanning modes provide the synthesis of nanostructured silicon dioxide particles or silicon nanoparticles. It was established that the principal angle of incidence for the nanostructured silicon is significantly reduced in comparison to the one for monocrystalline silicon wafers of 12–21°. Moreover, the essential difference between the values of the $\cos\Delta$ and $\mathrm{tg}\Psi$ for two mutually perpendicular directions in own plane of the cell when ones were measured at single taken angle of light incidence $\varphi = 53°$ was observed. Besides, for single cell it was found that the difference between the significances of the principal angle of incidence and ellipsometric parameter Ψ for two mutually perpendicular directions in own plane of the isolated cell of the nanostructured silicon is essential and equal to about 9° and more than 15° respectively. This means that the formed silicon nanostructures possess great optical anisotropy as a result of deformation influence of laser ablation and appearance of elastic stresses within the surface layer of the nanostructured silicon. The optical anisotropy was not found for silicon areas located between the cells of the nanostructured silicon.

Keywords: Nanostructured silicon · Laser ablation · Optical anisotropy

1 Introduction

Silicon is still widely used in micro- and nanoelectronics, sensor technology, biomedicine, etc. due to their thermodynamic, physical, chemical and semiconductor properties. Interest in the nanocomposite and nanostructured silicon based structures is caused by perspectivity of their use in the creation of silicon photonics devices [1] and nonvolatile memory devices [2]. Promising method to produce nanoparticles of different materials is laser ablation [3]. It is assumed that the processes of ejection (ablation) are purely thermal (thermal desorption, melting, overheating, explosive boiling)

© Springer Nature Switzerland AG 2019
G. Laukaitis (Ed.): INTER-ACADEMIA 2018, LNNS 53, pp. 66–72, 2019.
https://doi.org/10.1007/978-3-319-99834-3_9

and depend primarily on the value of the pulse energy density absorbed in the surface layer of the target. In particular, the transition from desorption to ablation and the quantity of ejected material is completely determined by radiation energy, the initial temperature of the target and the absorption coefficient [4].

One characteristic of ablation using high-power pulses of laser light is the formation of different kinds of particles (clusters, droplets or solid fragments). The ability to predict and control the ejection of particles and composition of clouds ablation products is essential for optimizing process parameters in many practical applications. Laser ablation is used the most widely in recent years in the preparation of nanocomposite materials for electronic devices.

In the work the results of ellipsometric studies of optical properties of monocrystalline and nanostructured silicon derived from the irradiation by fiber laser of the monocrystalline silicon wafer are presented.

2 Manufacturing Technology of Nanostructured Silicon and Experimental

Nanostructured silicon was formed on the surface of single-crystal silicon (100) orientation wafer with the natural oxide layer by laser ablation [3]. Irradiation was performed in an air atmosphere. The nanostructured silicon was formed by Yb-doped fiber laser which generates pulses with energies up to 1 mJ at a frequency of 1 MHz with a wavelength of 1060 nm. The power of laser irradiation on the surface of the monocrystalline silicon ranged from 100 mW to 1 W [5].

Scan modes of laser beam provide a synthesis of nanostructured particles of silicon dioxide or silicon nanoparticles. On the surface of silicon wafers the nanostructured layers in the form of separate cells were created. The schematic cellular structure in the form of rectangular patterns on the surface of silicon sample and optical microscopy image of the A3 cell are shown in Fig. 1(a, b). Wave structure as rowing along the short side of a rectangular cell is clearly visible on the surface. The individual cells are numbered from 1 to 20, and their placement on the surface of the sample is indicated by three rows oriented along the dashed lines A, B and C. Ellipsometric diagnostic was performed to certain cells which put concrete symbols such as position which is selected from the series A, B or C and cell number in the appropriate line. The size of the investigated cells 1–8 and 10–20 is of 15×7 mm^2 on the surface formed by action of ultrashort laser pulses onto surface structure. These areas are visually observed besides a cell at number 9 in Fig. 1 with the size of 7×7 mm^2.

Optical polarization measurements were carried out using laser ellipsometer LEF-3M-1 at a wavelength 632.8 nm [6] within the 15 divided cells on the wafer. The ellipsometric parameters of the samples such as a cosΔ and tgΨ were measured for two mutually perpendicular directions in own plane of the sample for characterization of optical anisotropy of some isolated cells of silicon samples at variation of the angle of light incidence. All angular dependencies of ellipsometric parameters cosΔ(φ) and tgΨ (φ) of the nanostructured silicon were analyzed and the principal angle of incidence (cosΔ = 0) and the angular positions of minimal value of tgΨ are obtained from these dependencies.

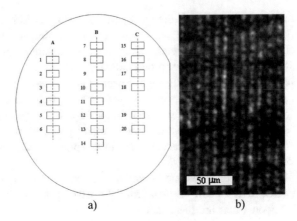

Fig. 1. The schematic cellular structure in the form of rectangular pattern on the surface of silicon (a) and optical microscopy image of the A3 cell (b)

3   Results and Discussion

At the first stage the study of the angular (due to changing angle φ) ellipsometric parameters $\cos\Delta$ and $\text{tg}\Psi$ were measured before laser treatment for a monocrystalline silicon wafer within 5 different areas at several angles of light incidence. The first area was chosen at the center of the unaffected silicon wafer, and other four areas were chosen at the edges of the plate. For each area the measurements of ellipsometric parameters were performed for two mutually perpendicular directions in own plane of the monocrystalline silicon wafer. The results of the experiment are given in Table 1. One can see that changes of the principle angle of incidence φ_p ($\cos\Delta = 0$) and the minimal value of the $\text{tg}\Psi$ not exceed the measurement errors namely $\delta\varphi_p = 0.01°$ and $\delta\text{tg}\Psi = 0.0013$. Consequently, the optical anisotropy was not found for monocrystalline silicon wafer.

Table 1. The ellipsometric parameters φ_p and the minimal value of the $\text{tg}\Psi$ for different five areas on unaffected monocrystalline silicon wafer for two angular directions in own plane of the sample $\alpha = 0°$ and $\alpha = 90°$

Area	Direction	φ_p	$\text{tg}\Psi_{min}$
1	$\alpha = 0°$	75.46	0.0250
	$\alpha = 90°$	75.46	0.0250
2	$\alpha = 0°$	75.46	0.0270
	$\alpha = 90°$	75.45	0.0280
3	$\alpha = 0°$	75.45	0.0267
	$\alpha = 90°$	75.46	0.0270
4	$\alpha = 0°$	75.44	0.0262
	$\alpha = 90°$	75.45	0.0266
5	$\alpha = 0°$	75.46	0.0261
	$\alpha = 90°$	75.46	0.0272

The ellipsometric parameters cosΔ and tgΨ depending on the angle of incidence of light on the sample were determinated for cell B10 of nanostructured silicon. The angle dependences of these two parameters for B10 cell of silicon sample measured under the condition where the direction B was perpendicular to the p-plane of the sample (curves 1, azimuthal angle α = 90°) or coincided with it (curves 2, azimuthal angle α = 0°) are presented in Fig. 2(a, b).

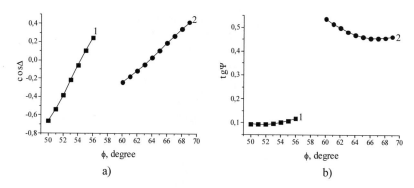

Fig. 2. Dependencies of cosΔ(φ) (a) and tgΨ(φ) (b) for B10 cell of the irradiated monocrystalline silicon sample at orientation of the long side of the cell parallel to its p-plane, α = 90° (1), and perpendicular to p-plane, α = 0°(2)

One can see that these dependencies cosΔ(φ) and tgΨ(φ) are significantly different for two cell orientations relatively to the p-plane. The principle angle of incidence (cosΔ = 0) changes almost 9° and the variation of the angular (due to φ changing) position for the minimal value of tgΨ shifts by more than Δφ = 15°. This means that the diagnosed B10 cell due to formed therein surface structure after the action of the powerful laser possesses optical anisotropy namely a measure which may be taken into account for its characterization is a difference in the angular position of the principle angle of incidence as well as variation Δφ of angular position for the minimal values of tgΨ for such orientations of cell as α = 0° (the short side of the B10 rectangle is placed along the p-plane) and α = 90° (the long side of the B10 rectangle is placed along the p-plane) in own plane of the silicon sample.

The values of the ellipsometric parameters cosΔ and tgΨ for different rectangular cells (Fig. 1a) on the sample surface at the angle of incidence φ = 53° for two azimuthal directions α = 0° and α = 90° in own plane of the sample are presented in Table 2. From these data (Table 2) it is seen that for each cell from all these cells within three rows A, B and C (Fig. 1) the values of tgΨ at α = 0° are almost in one order more than those recorded at azimuthal position α = 90°. Moreover, for the all cells at α = 0° and α = 90° the values of cosΔ are changed and possess to the same dependencies namely the absolute value of this parameter is 3.5–4 times higher in α = 0° than ones at α = 90°.

Some deviation for ellipsometric parameters is observed only for cell B9 which is primarily connected with sensing for half size of the cell subjected to laser treatment. In

Table 2. Ellipsometric parameters cosΔ and tgΨ for different rectangular cells on the silicon sample surface at the angle of incidence $\varphi = 53°$ for two azimuthal directions $\alpha = 0°$ and $\alpha = 90°$ in own plane of the sample

No	Direction	cosΔ	tgΨ
A1	$\alpha = 90°$	−0.1811	0.0938
	$\alpha = 0°$	−0.4652	0.8690
A2	$\alpha = 90°$	−0.1338	0.0944
	$\alpha = 0°$	−0.4813	0.8721
A3	$\alpha = 90°$	−0.1680	0.0917
	$\alpha = 0°$	−0.4675	0.8408
A4	$\alpha = 90°$	−0.2210	0.0964
	$\alpha = 0°$	−0.5113	0.9328
A5	$\alpha = 90°$	−0.2448	0.1198
	$\alpha = 0°$	−0.5803	0.9601
A6	$\alpha = 90°$	−0.2300	0.1113
	$\alpha = 0°$	−0.5676	0.8801
B7	$\alpha = 90°$	−0.2771	0.0912
	$\alpha = 0°$	−0.5061	0.8057
B8	$\alpha = 90°$	−0.2349	0.0948
	$\alpha = 0°$	−0.5240	0.8366
B9	$\alpha = 90°$	−0.9804	0.0838
	$\alpha = 0°$	0.1469	0.6132
B10	$\alpha = 90°$	−0.2037	0.0960
	$\alpha = 0°$	−0.5048	0.8468
B11	$\alpha = 90°$	−0.2301	0.0986
	$\alpha = 0°$	−0.5218	0.9094
B12	$\alpha = 90°$	−0.2148	0.0979
	$\alpha = 0°$	−0.5103	0.9182
B13	$\alpha = 90°$	−0.2281	0.0947
	$\alpha = 0°$	−0.5071	0.8655
C17	$\alpha = 90°$	−0.1740	0.0973
	$\alpha = 0°$	−0.5235	0.8632
C18	$\alpha = 90°$	−0.2094	0.1111
	$\alpha = 0°$	−0.5686	0.8790

this case on the surface is not visually observed wavy structure as rowing along the short side of this rectangular cell (B9) compared to clearly visible such rowing in all other cells. These data have confirmed, firstly, high uniformity of the optical properties of areas formed on the silicon surface by laser action, and secondly, these properties are anisotropic for each cell because of the values cosΔ and tgΨ for two mutually perpendicular directions of orientation of the sample in its own plane are differed significantly.

This research in some details is similar to our previous one for amorphous metal alloys. Their optical properties formed by skin layer were also anisotropic [7] indicating the presence of the strain and the appearance of elastic stresses inside the layer surface within the sample areas subjected to laser processing too. Such stress distribution within the skin layer probed by polarized light affects the atomic and electronic structure of this layer and variations in its optical response to the light excitation are clearly recorded by ellipsometric diagnostics in 2 perpendicular azimuthal directions. Then similar situation is observed for the silicon surface subjected to laser treatment.

The ellipsometric parameters cosΔ and tgΨ were also measured for silicon areas located between the cells 3, 4, 10 and 11 and 9, 10, 17 and 18 of the nanostructured silicon sample respectively. The experimental results for areas located between the cells 3, 4, 10 and 11 are presented in Fig. 3(a, b).

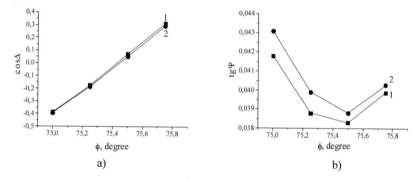

a)   b)

Fig. 3. Dependencies of cosΔ(φ) (a) and tgΨ(φ) (b) for areas located between the cells 3, 4, 10 and 11 of the irradiated monocrystalline silicon sample at orientation of the long side of the cells parallel to its p-plane, $\alpha = 90°$ (1), and perpendicular to p-plane, $\alpha = 0°$ (2)

The similar result is obseved for areas located between the cells 9, 10, 17 and 18. It is seen that optical anisotropy was not found for areas located between the cells of the nanostructured single silicon specimen.

4   Conclusions

The ellipsometric method is strongly sensitive to detect optical anisotropy of the surface structure of nanocrystalline silicon sample formed by laser processing. It was found that ellipsometric parameters obtained in two mutually perpendicular directions in own plane of the sample relatively to its p-direction are significantly different both in the case of multi-angle measurements and due to measuring cosΔ and tgΨ at single light incidence angle for various azimuthal sample positions in its own plane. Thus, such behavior of optical properties is a consequence of the presence of the deformation and elastic stresses formed on the surface layer of the single silicon sample within areas subjected to laser processing.

References

1. Fernandez, B., Garrido, L., Garcıa, C., et al.: Influence of average size and interface passivation on the spectral emission of Si nanocrystals embedded in SiO_2. J. Appl. Phys. **91**, 798–807 (2002)
2. Tiwari, S., Rana, F., Chan K., et al.: Volatile and nonvolatile memories in silicon with nanocrystal storage. IEEE Int. Electron Devices Meet. Tech. Dig. **42**, 521–524 (1995)
3. Veyko, V., Libenson, M., Chervyakov, G., Yakovlev, E.: Interaction of laser radiation with substance. Phismatlit, Moskow (2008). (in Russian)
4. Zhigilei, L., Leveugle, E., Garrison, B., et al.: Computer simulations of laser ablation of molecular substrates. Chem. Rev. **103**(2), 321–347 (2003)
5. Öktem, B., Pavlov, I., Ilday, S., et al.: Nonlinear laser lithography for indefinitely large-area nanostructuring with femtosecond pulses. Nat. Photonics **7**, 897–901 (2013)
6. Azzam, R., Bashara, N.: Ellipsometry and Polarized Light. Elsevier Science Ltd, Amsterdam (1987)
7. Poperenko, L., Kudryavtsev, Y., Stashchuk, V., Li, Y.: Optics of Metal Structures. Kyiv University, Kyiv (2013). (in Ukraine)

XPS Study of the In/CdTe Interface Modified by Nanosecond Laser Irradiation

Kateryna Zelenska[1,2(✉)], Volodymyr Gnatyuk[1,3] ⓘ,
Hideki Nakajima[4], Wanichaya Mekprasart[5], and Wisanu Pecharapa[5]

[1] Research Institute of Electronics, Shizuoka University, 3-5-1 Johoku, Naka-ku,
Hamamatsu 432-8011, Japan
czelenska@gmail.com
[2] Faculty of Physics, Taras Shevchenko National University of Kyiv,
Prospekt Akademika Glushkova 4, Kiev 03127, Ukraine
[3] V.E. Lashkaryov Institute of Semiconductor Physics, National Academy of
Sciences of Ukraine, Prospekt Nauky 41, Kiev 03028, Ukraine
[4] Synchrotron Light Research Institute, 111 University Avenue, Muang District,
Nakhon Ratchasima 30000, Thailand
[5] College of Nanotechnology, King Mongkut's Institute of Technology
Ladkrabang, 1 Thanon Chalong Krung, Ladkrabang, Bangkok 10520, Thailand

Abstract. The X-ray photoelectron spectroscopy (XPS) with 650 eV synchrotron radiation was employed to study the In/CdTe diode detector structures formed by laser-induced doping and subjected to multiple Ar-ion etching from the In-coated side. The dependences of the peak areas in the high-resolution Cd 3d and In 3d XPS spectra on the etching number demonstrated the presence of Cd atoms in the In film and In dopant atoms in the CdTe near the In/CdTe interface. This was attributed to ultrafast mutual diffusion under laser action. The XPS results were direct evidence of incorporation of In dopant atoms into the surface region of the CdTe crystal and penetration of Cd atoms into the In film.

Keywords: CdTe crystals · Laser irradiation · Doping · Diffusion
XPS spectra

1 Introduction

High-resistivity THM-grown CdTe semiconductor is known as industrial material for high efficiency good energy resolution room-temperature X/γ-ray detectors. Detector-grade Cl-compensated (111) oriented p-like CdTe single crystals, manufactured by Acrorad Co. Ltd., are quite uniform with a decreased number of native point and extended defects, low amount of accidental impurities and excellent electrical characteristics [1–3]. Such crystals have been generally used for fabrication of diode-type detectors implemented as structures either with a Schottky barrier [1–7] or p-n junction [7–13]. In this case, processing of the CdTe surface plays a key role in detector performance. Mechanical and chemical polishing, etching with Br-methanol solutions,

© Springer Nature Switzerland AG 2019
G. Laukaitis (Ed.): INTER-ACADEMIA 2018, LNNS 53, pp. 73–79, 2019.
https://doi.org/10.1007/978-3-319-99834-3_10

inorganic acids, ion bombardment or laser radiation are usually involved in semiconductor surface treatments before the formation of surface barrier structures [1–13].

Employing irradiation of p-CdTe crystals pre-coated with an In dopant film by nanosecond laser pulses, it is possible to incorporate and activate In atoms (donors) with high concentration in the thin CdTe surface region and thus, to form a shallow and sharp p-n junction [7–13]. Highly non-equilibrium and non-stationary processes under nanosecond laser action resulted extremely fast diffusion of intrinsic atoms and impurities, formation, transformation and redistribution of point defects, etc. [9–13].

Despite the recent advances in laser-induced doping and fabrication of CdTe-based diodes with a p-n junction developed using chemical etching and nanosecond laser irradiation, many issues still remain unclear or unsolved, particularly with regard to stoichiometry and elemental composition in the CdTe surface layer, dopant profile in the thin CdTe region near the metal-semiconductor interface, concentration and distribution of impurities, component composition and homogeneity of the laser-modified layers, etc. [10–13].

In this study, X-ray photoelectron spectroscopy (XPS) was employed to get structural and chemical information about the surface and thin superficial layers of the In/CdTe diode structure formed by laser-induced doping [14]. XPS is a unique probe to determine stoichiometry and identify contaminations, dopants, accidental impurities, their distribution with surface layer thickness and other data which can be obtained by analysis of the electron behavior based on the band structure [15–18].

XPS spectra of the In/CdTe diode structure were repeatedly taken to characterize thin layers of the sample in the buried regions (in the In film, In-CdTe interface and CdTe surface area) as the sample was exposed by multiple ion etching.

2   Experimental Details and Analysis

The In/CdTe/Au diode structures were formed by the developed techniques of laser-induced solid-phase doping of the thin surface region of high resistivity p-like CdTe single crystals pre-coated with an In dopant film [7–12]. Detector-grade Cl-compensated (111) oriented p-like CdTe semiconductor crystals with the resistivity of ~ 109 Ω·cm produced by Acrorad Co., Ltd. were used [1–3]. Parallelepiped-like wafers with the size of $5 \times 5 \times 0.5$ mm^3 preliminary polished by the manufacturer were subjected to preliminary surface processing for cleaning and removing a disordered surface layer: washing in acetone and methanol, polishing etching in a Br-methanol solution and following rinsing in methanol. A relatively thick (300–500 nm) In dopant film was deposited on the CdTe(111)B surface by vacuum evaporation.

The surface area of the CdTe crystal pre-coated with an In film was entirely and uniformly irradiated with single nanosecond pulses of a KrF excimer laser ($\lambda = 248$ nm; $\tau = 20$ ns) in water at the depth of ~ 3–5 mm (Fig. 1, inset). The deposited In film was not completely evaporated under multiple laser irradiation and it served as both an n-type dopant source during laser irradiation and also as an electrode after laser-induced doping was performed [8–12].

The CdTe (111) crystals with a pre-deposited In dopant film (~ 400 nm), subjected to irradiation with nanosecond laser pulses, were studied by XPS. It was supposed that

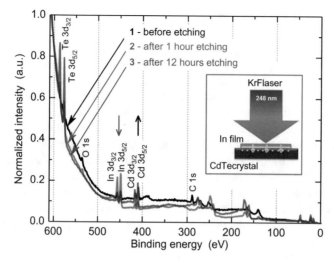

Fig. 1. XPS wide scan (survey) spectra of the In/CdTe structure, subjected to irradiation with KrF excimer laser pulses ($\lambda = 248$ nm; $\tau = 20$ ns) and long-term storage in air (curve 1), and then after etching with Ar ions during 1 h (curve 2) and 12 h (curve 3). Insert shows the procedure of laser-induced doping employing KrF excimer laser pulses

laser irradiation of the film resulted in In atom (dopant) incorporation due to generation of stress and shock waves, barodiffusion and thus, formation of a heavily doped CdTe layer near the metal-semiconductor interface [9–12]. This layer was studied with indirect methods using electrical and photoelectric measurements to reveal the In doped n-CdTe surface layer under the In electrode [11, 12]. The corresponding simulations of the laser-stimulated processes (superfast heating, stress and shock wave formation, enhanced diffusion, rapid mass transfer) were also performed [8–10]. The mechanisms of laser-induced transformation of the point defect structure and solid-phase doping of the CdTe surface region were discussed [10, 12, 13].

However, direct techniques have not been applied to study the laser-modified layers in the fabricated In/CdTe diode structures. Therefore, it was interesting and practically useful to study the surface and subsurface states in such structures, in particular to reveal In dopant atoms, determine the distribution of impurities with depth, analyze the stoichiometry and elemental composition of theses layers. As known, XPS is a very suitable and powerful technique for these tasks [15–18].

The XPS spectra of In/CdTe diode structures were obtained under irradiation from the In-coated side with a high-spatial resolution X-ray beam of 650 eV using a synchrotron excitation source at the Synchrotron Light Research Institute (Nakhon Ratchasima, Thailand) [14]. The In film was repeatedly etched 12 times with Ar plasma using a gun operating at 3.0 keV and 5.7 mA and producing a specimen current of approximately 3 μA. The XPS wide scan spectra study and precise measurements of the selected peaks for the etched samples were carried out in the ultra-high vacuum ($<2 \cdot 10^{-10}$ mbar) after every hour of the Ar-ion sputtering procedure.

3 Results and Discussion

Capabilities of XPS for compositional analysis (atomic abundance of elements in metals, semiconductors and insulators), chemical analysis (identification of bonds between specific atoms in the surface region of different materials), through-thickness analysis (depth profiles characterizing buried regions, using exposition by ion etching) have made the wide possibilities to study and obtain the data for purposeful modification of surface layers of CdTe crystals, In/CdTe interface and deeper CdTe regions subjected to nanosecond laser irradiation and stimulated doping [15–18].

Figure 1 shows the XPS wide scan spectra of the In/CdTe diode surface before (curve 1) and after Ar-ion etching from the In-coated side with different time (curve 2 and curve 3), demonstrating the features of In, Cd and Te as result of measurement of electron binding energies. The core level, Auger and valence electron peaks of In, Cd, Te, O and C were used for the detailed analysis of the component composition of the area adjacent to the In/CdTe interface.

As seen from Fig. 1, the Te 3d peaks are absent in the initial In/CdTe structure before etching (curve 1). Then, they appear after 1 h etching and increase with next 1 h etchings, totally 12 h of etching (curve 2 and curve 3). In the behavior of the In 3d and Cd 3d peaks with a spin orbital splitting, a decrease of the In 3d and increase in the Cd 3d peak intensities after each 1 h Ar-ion etching are observed (the sample was etched 12 times during 1 h) (Fig. 1, curve 2 and curve 3).

The peaks related to contaminations such as oxygen (O 1s peak) and carbon (C 1s peak) are manifested in the XPS spectrum of the initial In/CdTe structure (Fig. 1, curve 2, curve 1). These features tend to disappear after 1–2 h of etching (Fig. 1, curve 2) and not observed after multiple etching (Fig. 1, curve 3). This was evidence that contaminations were only on the surface and removed after first ion etchings.

Figure 2 (curve 1 and curve 2) demonstrates the peak areas calculated from the high resolution XPS spectra for the Cd 3d and In 3d, respectively in dependence of the number of etching. The minimum value of the In 3d peak before etching can be attributed to presence of accidental impurities on the surface. Based on the obtained XPS data, the distribution of In and Cd atoms with the depth of the In/CdTe structure was estimated. Cd atoms were revealed in the In film superficial layer which was non-contiguous to the CdTe surface. The obtained results can be explained by the laser-stimulated processes of ultrafast heating, stress generation and as result ultrafast mutual diffusion of Cd atoms from the semiconductor into the adjoined metal film and also by penetration of In dopant atoms into the CdTe crystal surface region.

The phenomena of abnormally fast laser-stimulated mass transfer and barodiffusion under stress and shock waves action were discussed as the possible mechanisms of laser-induced doping and migration of In and Cd atoms in the In/CdTe structure [8–10]. The experimental direct evidence of incorporation of In dopant atoms into the surface region of the CdTe crystal and formation of the impurity enriched surface region have been obtained by the XPS spectra measurement.

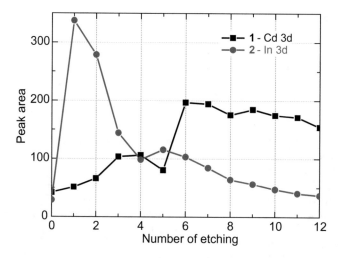

Fig. 2. Dependences of the peak areas in the high resolution Cd 3d (curve 1) and In 3d (curve 2) XPS spectra of the In/CdTe diode structure subjected to multiple Ar-ion etching (during 1 h each) on the number of etching

4 Conclusion

Thus, based on the XPS spectra measured for In/CdTe structures, subjected to irradiation with nanosecond laser pulses and following multiple etching with Ar plasma, we can conclude that In impurity (dopant) atoms have been directly revealed in the surface region. Moreover, Cd atoms were found in the In film that was attributed to ultrafast laser-stimulated mutual diffusion. The further investigation and analysis of the obtained XPS spectra will give detailed information about the chemical composition and atomic structure of the surface CdTe:In nano-layer near the In/CdTe interface after laser-induced doping with In impurity and it will allow us to clarify the actual mechanisms of laser-stimulated processes under doping to find the optimal conditions and parameters of the employed technique.

Acknowledgements. This research was supported by the following international projects and programs: (1) "A sensor network for the localization and identification of radiation sources" (SENERA, grant SfP-984705) of the NATO Science for Peace and Security Programme; (2) the Academic Melting Pot Project at King Mongkut's Institute of Technology Ladkrabang (KMITL), Bangkok, Thailand; (3) "Development of Cd(Zn)Te-based X/gamma-ray detectors with high resolution for security and diagnostics instruments" (grant No. 2035) of the 2018 Cooperative Research at Research Center of Biomedical Engineering, Japan.

References

1. Funaki, M., Ando, Y., Jinnai, R., Tachibana, A., Ohno, R.: Development of CdTe detectors in Acrorad. In: International Workshop on Semiconductor PET, pp. 1–8 (2007). https://directconversion.com/wp-content/uploads/2017/05/Acrorad_Development_of_CdTe_detectors.pdf
2. Shiraki, H., Funaki, M., Ando, Y., Tachibana, A., Kominami, S., Ohno, R.: THM growth and characterization of 100 mm diameter CdTe single crystals. IEEE Trans. Nucl. Sci. **56**(4), 1717–1723 (2009)
3. Shiraki, H., Funaki, M., Ando, Y., Kominami, S., Amemiya, K., Ohno, R.: Improvement of the productivity in the THM growth of CdTe single crystal as nuclear radiation detector. IEEE Trans. Nucl. Sci. **57**(1), 395–399 (2010)
4. Kosyachenko, L.A., Aoki, T., Lambropoulos, C.P., Gnatyuk, V.A., Sklyarchuk, V.M., Maslyanchuk, O.L., Grushko, E.V., Sklyarchuk, O.F., Koike, A.: High energy resolution CdTe Schottky diode γ-ray detectors. IEEE Trans. Nucl. Sci. **60**(4), 2845–2852 (2013)
5. Maslyanchuk, O.L., Solovan, M.M., Kulchynsky, V.V., Brus, V.V., Maryanchuk, P.D., Fodchuk, I.M., Gnatyuk, V.A., Aoki, T., Potiriadis, C., Kaissas, Y.: Capabilities of CdTe-based detectors with MoO$_x$ contacts for detection of X- and γ-ray radiation. IEEE Trans. Nucl. Sci. **64**(5), 1168–1172 (2017)
6. Sklyarchuk, V.M., Gnatyuk, V.A., Pecharapa, W.: Low leakage current Ni/CdZnTe/In diodes for X/γ-ray detectors. Nucl. Instrum. Methods Phys. Res. A **879**, 101–105 (2018)
7. Lambropoulos, C.P., Aoki, T., Crocco, J., Dieguez, E., Disch, C., Fauler, A., Fiederle, M., Hatzistratis, D.S., Gnatyuk, V.A., Karafasoulis, K., Kosyachenko, L.A., Levytskyi, S.N., Loukas, D., Maslyanchuk, O.L., Medvids, A., Orphanoudakis, T., Papadakis, I., Papadimitriou, A., Potiriadis, C., Schulman, T., Sklyarchuk, V.M., Spartiotis, K., Theodoratos, G., Vlasenko, O.I., Zachariadou, K., Zervakis, M.: The COCAE detector: an instrument for localization - identification of radioactive sources. IEEE Trans. Nucl. Sci. **58**(5), 2363–2370 (2011)
8. Gnatyuk, V.A., Dubov, V.L., Fomin, D.V., Seteikin, A.Yu., Aoki, T.: Temperature fields in the In/CdTe structure under laser-induced doping in liquid. In: Recent Advances in Technology Research and Education, Inter-Academia 2017. Advances in Intelligent Systems and Computing, vol. 660, pp. 87–95 (2017)
9. Gnatyuk, V.A., Aoki, T., Hatanaka, Y.: Laser-induced shock wave stimulated doping of CdTe crystals. Appl. Phys. Lett. **88**(24), 242111-1-3 (2006)
10. Veleshchuk, V.P., Baidullaeva, A., Vlasenko, A.I., Gnatyuk, V.A., Dauletmuratov, B.K., Levitskii, S.N., Lyashenko, O.V., Aoki, T.: Mass transfer of indium in the In-CdTe structure under nanosecond laser irradiation. Phys. Solid State **52**(3), 469–476 (2010)
11. Gnatyuk, V.A., Levytskyi, S.N., Vlasenko, O.I., Aoki, T.: Electrical properties of CdTe-based structures with an *n*-layer formed by laser-induced doping. J. Adv. Res. Phys. **2**(2), 021103-1-4 (2011)
12. Gnatyuk, V.A., Levytskyi, S.N., Vlasenko, O.I., Aoki, T.: Formation of doped nano-layers in CdTe semiconductor crystals by laser irradiation with nanosecond pulses. Thai J. Nanosci. Nanotechnol. **1**(2), 7–16 (2016)
13. Zelenska, K.S., Gnatyuk, D.V., Aoki, T.: Modification of the CdTe-In interface by irradiation with nanosecond laser pulses through the CdTe crystal. J. Laser Micro/Nanoeng. **10**(3), 298–303 (2015)

14. Nakajima, H., Tong-on, A., Sumano, N., Sittisard, K., Rattanasuporn, S., Euaruksakul, C., Supruangnet, R., Jearanaikoon, N., Photongkam, P., Chanlek, N., Songsiriritthigul, P.: Photoemission spectroscopy and photoemission electron microscopy beamline at the Siam Photon Laboratory. J. Phys. Conf. Ser. **425**(13), 132020-1-4 (2013)

15. Waag, A., Wu, Y.S., Bicknell-Tassius, R.N., Landwehr, G.: Investigation of CdTe surfaces by x-ray photoelectron spectroscopy. Appl. Phys. Lett. **54**(26), 2662–2664 (1989)

16. Duszak, R., Tatarenko, S., Cibert, J., Saminadayar, K., Deshayes, C.: (111) CdTe surface structure: a study by reflection high energy electron diffraction, x-ray photoelectron spectroscopy, and x-ray photoelectron diffraction. J. Vac. Sci. Technol. A **9**(6), 3025–3030 (1991)

17. Bassani, F., Tatarenko, S., Saminadayar, K., Magnea, N., Cox, R.T., Tardot, A., Grattepain, C.: Indium doping of CdTe and $Cd_{1-x}Zn_xTe$ by molecularbeam epitaxy: uniformly and planardoped layers, quantum wells, and superlattices. J. Appl. Phys. **72**(7), 2927–2939 (1992)

18. Wang, X., Campbell, C., Zhang, Y.H., Nemanich, R.J.: Band alignment at the CdTe/InSb (001) heterointerface. J. Vac. Sci. Technol. A Vac. Surf. Films **36**(3), 031101-5 (2018)

Mechanical Properties of Cellular Structures with Schwartz Primitive Topology

Maxim M. Sychov[1,2(✉)], Lev A. Lebedev[1,2], Alexei A. Evstratov[3], Arnaud Regazzi[3], and Jose-Marie Lopez-Cuesta[3]

[1] Institute of Silicate Chemistry, Russian Academy of Sciences,
St. Petersburg, Russia
msychov@yahoo.com
[2] St. Petersburg State Institute of Technology (Technical University),
St. Petersburg, Russia
[3] IMT, Ales, France

Abstract. Cellular structures with triply periodic minimal surface (TPMS) topology are studied in current research work. Previously, it was shown that materials with TPMS topology have a potential as a lightweight material in energy-absorbing systems that can be used in various applications such as protective layer landing pods of spacecraft. In this work Schwartz primitive structure was chosen to study the influence of scale factor, unit number on mechanical properties. Samples were made by selective laser sintering process of polyamide powder and tested on Walter+bai ag LFM − 400 kN machine.

Keywords: Additive manufacturing · TPMS · Minimal surface
Mechanical properties

1 Introduction

In recent years lightweight materials attracts a lot of interest for construction and functional structures, because of energy efficiency provided by smaller weight of vehicles, aircrafts, spacecrafts, etc. One route to fabricate materials with small densities and high rigidity and strength is utilization of composites made of polymer matrix and reinforcement material in dispersed, fiber or sheet form. The other route is structural where not only composition determines mechanical properties, but also how this material arranged in space. There are statistical or random arrangement that can be found in foams such as foamed concrete, and ordered or cellular arrangement such as some types of cardboard and ribs in air or underwater vehicles hull [1–3].

One of the most widely used ordered structures is honeycomb-like structures. Such structures are used as rigid and durable infill for energy absorbers in car bumpers, space pods, ship protection and armor [4–6].

Design of cellular structures determines its properties and application, it can be produced in a number of basic shapes like tubes, spheres, frameworks, etc. or specifically generated with topology optimization software to produce so called bionic structures in case if they resemble or mimic biological objects [7, 8].

© Springer Nature Switzerland AG 2019
G. Laukaitis (Ed.): INTER-ACADEMIA 2018, LNNS 53, pp. 80–86, 2019.
https://doi.org/10.1007/978-3-319-99834-3_11

There are lots of structures, with remarkable properties, that can be found in nature. Some of them are periodic frameworks with specific topology, which enables low density (high porosity) and high strength at the same time. This sort of materials can be found in bug shells and butterfly wing scales, this structure, due to small period of "lattice", are responsible for structural colors of the insects as well. Topology of these type bionic materials is similar to Triply Periodic Minimal Surfaces (TPMS) [9, 10].

In works of V. Ya. Shevchenko it was shown that materials with TPMS topology may be perspective for utilization I conditions of extreme loads [11, 12].

In our previous work [13] plastic samples with several TPMS topologies were studied. It was shown that their energy absorbing properties are suitable for shock absorbing shell of space landing pods. In current research Schwartz Primitive topology is chosen as model structure to investigate influence of size and scale effects on mechanical properties of samples.

2 Experimental

2.1 Materials and Methods

Software. To make 3D models with TPMS topology in "stl" format there was used a script that allows to derive topology from mathematical formula. The script was made in Rhinoceros 3D modeling software with Grasshopper-graphical algorithm editor and a pack of plugins such as "weaverbird" to get a proper instrument and functions. Generation of G-code was carried out in PSW slicing program.

Materials and Equipment. Samples were printed using EOS Formiga 110 SLS 3D-printer with polyamide 12 powder. Mechanical properties were measured on a Breaking Machine Walter+bai ag LFM − 400 kN at the 26 °C temperature and 5 mm/min loading rate.

3 Results and Discussion

All samples were shellular. Equation below, represents formula of Schwartz Primitive surface, where t determines position of surface in unit cell and volumes of pore systems. Parameter t in all samples equals 0. For all series three samples of each type were printed and measured to provide reproducibility.

$$\cos(x) + \cos(y) + \cos(z) = t \tag{1}$$

Scale Change Series. There were tested four sample types with different scale factor with 27 unit cells (3 × 3 × 3 culls cubes) as shown on the Fig. 1, all relative sizes were constant. Sizes were scaled as double, triple and quadruple from the smallest sample, it gave sample size L as 15 mm, 30 mm, 45 mm, 60 mm with wall thickness 0.4 mm, 0.8 mm, 1.2 mm, 1.6 mm respectively.

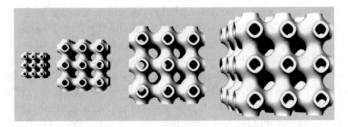

Fig. 1. Scale change series render, with side length L from left to right 15 mm, 30 mm, 45 mm, 60 mm

Stress-strain curves show plastic type deformation which is typical for nylon and negligible scale factor effect for all samples except smallest one that is softer in case of thin walls, and its curve lowered in contrast to others due to small absolute force values, that will result in high error value (Fig. 2).

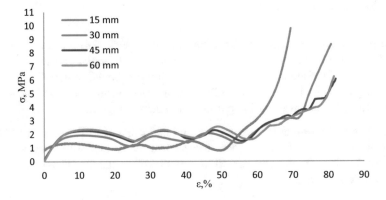

Fig. 2. Compression curves of scale change series

Stress-strain curves have three distinct regions: elastic ($\varepsilon \sim 0\%...10\%$), the condensing-plastic ($\varepsilon \sim 10\%...50\%$), and the densification ($\varepsilon > 50\%$). Strengths values from side length curve (Fig. 3) shows, that mechanical strength raised from 1.3 to 2.3 MPa with increase of sample size from 15 to 60 mm. There is an obvious power low in strength increase tendency with exponent value 0.4. Samples showed high plasticity under compression without destruction, they folding like a spring instead and thus are quite suitable for energy absorbing applications.

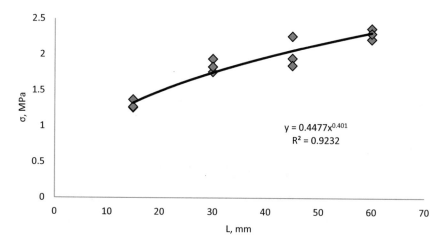

Fig. 3. Mechanical strength depending from sample size

Unit Amount Change Series. In this series unit number on strength value was tested. In all samples unit parameter was constant, with size of $10 \times 10 \times 10$ mm on wall thickness 0.8 mm. Number of units in rib (N) changed from 2 ($2 \times 2 \times 2$ units cube) to 6 ($6 \times 6 \times 6$ units cube), gives 5 types in total (Fig. 4).

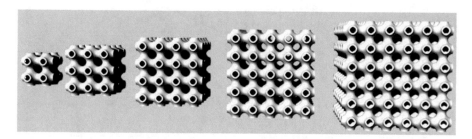

Fig. 4. Unit amount change series render, with side length from left to right 20 mm, 30 mm, 40 mm, 50 mm, 60 mm

Regarding compression strain curves (Fig. 5) one can see that all curves have first maximum (strength) and start of densification in the similar regions of deformation at 10% and 70% respectively. But number of peaks in between of this ε values is different and equals to the number of units in rib, thus providing smoother damping.

At the same time, exponent value for the size dependence of strength is lower in this case, Fig. 6. Position of first maximum on stress-strain curve gradually increases with L rise but there is a change in the slope indicating change in character of sample destruction under the load, Fig. 7.

To estimate prospects of usage of structures with the TPMS topology as energy absorbers, the damping characteristics of the samples were analyzed using specific impact absorption energy (A) value calculated according to equation:

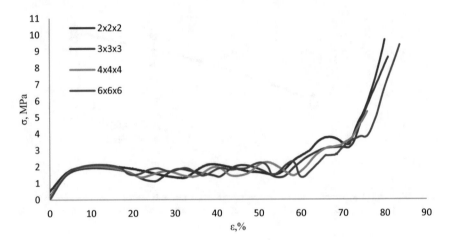

Fig. 5. Compression curves of unit amount change series

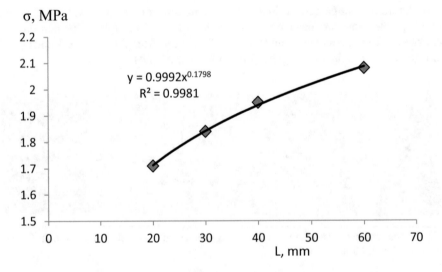

Fig. 6. Mechanical strength depending from sample size

$$A = \int_0^{\varepsilon_{max}} \sigma d\varepsilon \qquad (1)$$

Values of A are shown in the Fig. 8. As a maximum value of strain we used 50%. One can see that A values have similar trend of increase with sample length rise as it was observed for sample strength.

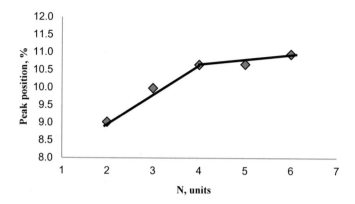

Fig. 7. Position of first maximum on stress-strain curve

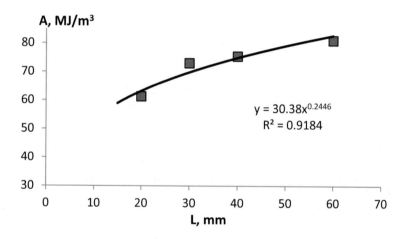

Fig. 8. Specific impact absorption energy

4   Conclusion

In this research work effect of scale factor and unit number on mechanical properties of hollow shellular polymer samples with Schwarts primitive topology were studied. It was shown that frameworks of specific TPMS types with large number of unit cells can be used as energy absorbing materials.

Acknowledgments. The work was financially supported by the Russian Science Foundation (project no. 17-13-01382).

References

1. Slutsker, A.B., Sinani, V.I., Betekhtin, A.A., Kozhushko, A.G., Kadomtsev Ordyan, S.S.: Effect of microporosity on the strength properties of SiC ceramics. Phys. Solid State **50**(8), 1395–1401 (2008)
2. Chu, J., Engelbrecht, S., Graf, G., Rosen, D.W.: A comparison of synthesis methods for cellular structures with application to additive manufacturing. Rapid Prototyp. J. **16**(4), 275–283 (2010)
3. Rosen, David W.: Computer-aided design for additive manufacturing of cellular structures. Comput. Aided Des. Appl. **4**(5), 585–594 (2007)
4. Alyoshin, V.F., Kolobov, AYu., Makarov, V.P., Petrov, Yu.A.: Landing devices of space vehicles (SC) on the basis of foams and sotoblocks. Sci. Educ. Sci. Tech. Publ. **77**(4), 1–7 (2010)
5. Markov, V.A., Pusev, V.I., Selivanov, V.V.: Questions of application of highly porous metals and honeycomb structures for protection against shock wave loads. Quest. Defensive Eng. **7–8**, 54–62 (2012)
6. Paik, J.K., Thayamballi, A.K., Kim, G.S.: The strength characteristics of aluminum honeycomb sandwich panels. Thin-Walled Struct. **35**(3), 205–231 (1999)
7. Kushnir, O.A.: Comparison of the form of binary raster images based on skeletonization. Mach. Learn. Data Anal. **1**(3), 252–255 (2012)
8. Sun, J., Liu, C., Du, H., Tong, J.: Design of a bionic aviation material based on the microstructure of beetle's elytra. Int. J. Heat Mass Transf. **114**, 62–72 (2017)
9. Lord, E.A., Mackay, A.L., Ranganatan, S.: New Geometry for New Materials. Cambridge University Press, Cambridge (2006)
10. Andersson, S., Hyde, S.T., Larsson, K., Lidin, S.: Minimal surfaces and structures: from inorganic and metal crystals to cell membranes and biopolymers. Chem. Rev. **88**(1), 221–242 (1988)
11. Shevchenko, V.Ya., Sychev, M.M., Lapshin, A.E., Lebedev, L.A.: Glass Phys. Chem. **43**(6), 605–607 (2017)
12. Shevchenko, VYa., Sychev, M.M., Lopatin, A.E., Lebedev, L.A., Gruzdkov, A.A., Glezer, A.M.: Glass Phys. Chem. **43**(6), 608–610 (2017)
13. Sychov, M.M., Lebedev, L.A., Dyachenko, S.V., Nefedova, L.A.: Acta Astronaut. (2018, in press)

Plasma Physics

Effect of Damage Introduction and He Existence on D Retention in Tungsten by High Flux D Plasma Exposure

Yasuhisa Oya[1(✉)] [iD], Keisuke Azuma[1], Akihiro Togari[1],
Moeko Nakata[1], Qilai Zhou[1], Mingzhong Zhao[1], Tatsuya Kuwabara[2],
Noriyasu Ohno[2], Miyuki Yajima[3], Yuji Hatano[4],
and Takeshi Toyama[5]

[1] Shizuoka University, 836 Ohya, Suruga-ku, Shizuoka, Japan
oya.yasuhisa@shizuoka.ac.jp
[2] Nagoya University, 1247 Furo-cho, Chikusa-ku, Nagoya, Japan
[3] National Institute for Fusion Science, 322-6, Oroshi, Toki, Gifu, Japan
[4] University of Toyama, 3190, Gofuku, Toyama, Japan
[5] Tohoku University, 2145-2, Narita-cho, Oarai, Ibaraki, Japan

Abstract. Both of radiation-induced damages and helium (He) existence effects on deuterium (D) retention in tungsten (W) by D plasma exposure were evaluated using high flux divertor plasma exposure device, called Compact Divertor Plasma Simulator (CDPS). The results were compared with 3 keV D_2^+ implanted W with low flux and fluence. The thermal desorption spectra were consisted of three desorption stages at 400, 600, 780 K. Comparing to the undamaged W, the D desorption stages were shifted towards higher temperature side and the values of D retention increased. It can be said that the formation of stable trapping sites by damage introduction enhances the D trapping in the damaged W. For He^+ irradiation, D desorption at lower temperature was enhanced, due to the formation of dense dislocation loops. In case of sequential Fe^{2+} and He^+ implantation, D desorption at higher temperature was reduced, comparing to that for only Fe^{2+} damaged W. These facts show that the accumulation of He near surface region reduces D diffusion toward bulk, leading to the reduction of D trapping by voids.

Keywords: Hydrogen isotopes · Irradiation damages · He bubble formation

1 Introduction

For the development of future fusion reactor, it is quite important to evaluate the fuel retention, especially tritium (^{3}H), in plasma facing walls to reduce hazard of radiation exposure and to use bred ^{3}H efficiently. Recently, tungsten (W) is thought to be the best plasma facing material due to high melting point, low sputtering yield and low hydrogen solubility [1, 2]. However, it is known that the radiation damages by energetic hydrogen isotope, helium (He) and neutron, which will be produced by DT fusion reaction, will be introduced into W leading to the formation of stable trapping sites [3–10]. Therefore, fuel retention behavior should be evaluated by simulating the actual fusion environment.

© Springer Nature Switzerland AG 2019
G. Laukaitis (Ed.): INTER-ACADEMIA 2018, LNNS 53, pp. 89–96, 2019.
https://doi.org/10.1007/978-3-319-99834-3_12

In our previous fundamental studies, deuterium (D) retention was clearly controlled by damage concentration and profiles in W [7–10]. In addition, the existence of He also changes the D retention behavior due to the formation of He bubbles near surface region [11–13]. It is thought that two possible mechanisms for the interaction of He with D in W. The first one is that the formation of He bubbles near the W surface reduces the D diffusion toward the bulk. The other is that He is trapped by radiation-induced damages to form He-vacancy complex, which may prevent the D trapping due to strong trapping of He in vacancy. It is quite important to evaluate damage introduction and He existence effects on D retention in W, plasma facing wall, for fusion safety analysis. Therefore, this study focuses on synergistic effects of radiation-induced damages and He on D retention in W using a high flux divertor plasma exposure device, called Compact Divertor Plasma Simulator (CDPS) at Tohoku University [14]. The results were compared with those at low D flux and fluence obtained using triple ion implantation system at Shizuoka University and discuss the D trapping behavior under co-existence of radiation-induced damages and He in W.

2   Experimental

The stress-relieved W with the size of 6 mm in diameter and 0.5 mm in thickness, was used as samples. The samples were polished as a mirror surface and annealed in vacuum at 1173 K for 30 min to remove surface impurities and diminish intrinsic defects. Thereafter, 6 MeV Fe^{2+} was irradiated into W to introduce the damages to 0.03–0.3 displacement per atom (dpa) at room temperature by 3 MV tandem accelerator, Takasaki Ion Accelerators for Advanced Radiation Application (TIARA) at National Institutes for Quantum and Radiological Science and Technology (QST). Based on Stopping and Range of Ions in Matter (SRIM) simulation, the implantation depth of 6 MeV Fe^{2+} was reached around 1.8 μm [15]. For He irradiation, 3 keV He^+ was irradiated at Shizuoka University up to the fluence of 1.0×10^{21} He^+ m^{-2}. For lower D_2^+ fluence experiment, 3 keV D_2^+ irradiation was performed at triple ion implantation system at Shizuoka University with the fluxes of 1.0×10^{18} D^+ m^{-2} s^{-1} and 3.0×10^{18} D^+ m^{-2} s^{-1} up to the fluences of 1.0×10^{22} D^+ m^{-2} and 3.0×10^{22} D^+ m^{-2}, respectively. The implantation depth for 3 keV He^+ and D_2^+ was estimated to be 50 nm. After the D_2^+ implantation, the sample was transferred in thermal desorption spectroscopy (TDS) chamber which was connected with the triple ion implantation system. The TDS was applied with the heating rate of 0.5 K s^{-1} from room temperature up to 1173 K to evaluate the D retention behavior in W.

In case of higher fluence experiment, D plasma exposure was performed using CDPS, which was installed in the radiation-controlled area at Tohoku University. Figure 1 shows the schematic view of CDPS. Typical D ion energy was ~ 100 eV and their flux was ~ 1.1×10^{21} m^{-2} s^{-1}. In the present study, the sample temperature was set to be kept at 373 K. The D plasma exposure was done up to the fluence of 5.0×10^{24} D^+ m^{-2}. After the plasma exposure, the sample was transferred to the infrared heater without air exposure and TDS was applied under the same conditions as the case of low fluence experiment. It was confirmed in advance using reference samples that almost the same desorption spectra were obtained using these two different TDS systems.

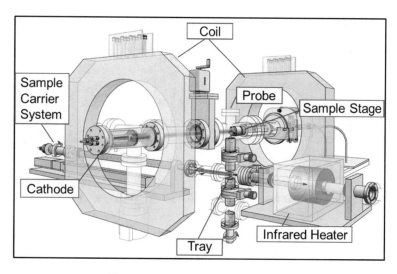

Fig. 1. Schematic view of CDPS [14]

3 Results and Discussion

Figure 2(a) and (b) show the D_2 TDS spectra for W with the damage level of 0.03 or 0.3 dpa with lower D fluence of 3.0×10^{22} D^+ m^{-2} and higher D fluence of 5.0×10^{24} D^+ m^{-2}. For lower fluence experiment, major D desorption for undamaged W was found at the lower temperature of 400 K, which corresponded to the desorption of D trapped by dislocation loops or near surface [7]. As the damage level increased, additional D desorption stages were found at higher temperature of 600 and 800 K, indicating that D was trapped by vacancies or voids. In 0.3 dpa damaged W, major D desorption was clearly shifted toward higher temperature of 800 K, although D desorption stage for 0.03 dpa damaged W was found at 600 K. It can be said that the accumulation of radiation-induced damages changes the D trapping states and dense vacancies form the voids in bulk W, where D is stably trapped. These behaviors were almost agreement with the results with higher fluence experiment of 5.0×10^{24} D^+ m^{-2} by D plasma exposure as shown in Fig. 2(b), but the desorption rate increased due to high fluence plasma exposure. Even for the undamaged W, the desorption stage was extended from 400 K to 650 K, indicating that the D was trapped by intrinsic damages like vacancies due to D plasma exposure with high flux, which would enhance the D diffusion toward the bulk, leading to the higher D retention. As the damage level was reached to 0.3 dpa, clear two desorption stages were found at 500 K and 800 K.

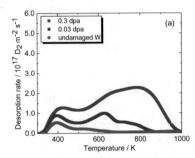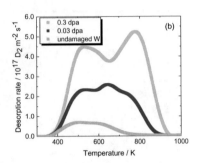

Fig. 2. D_2 TDS spectra for un-damaged W and damaged W with the damage concentrations of 0.03 and 0.3 dpa. (a) lower D fluence of 3.0×10^{22} D^+ m^{-2} (b) higher D fluence of 5.0×10^{24} D^+ m^{-2}

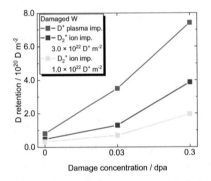

Fig. 3. Damage level dependence on D retention for W

Figure 3 summarizes the damage level dependence on D retention with various D fluences. For D plasma exposure with high D fluence, the D retention enhancement was clearly confirmed even in low damage level of 0.03 dpa.

To confirm the synergetic effect of radiation-induced damages and He on D retention in W, TDS experiments with same procedure were performed for only He^+ implanted W and Fe^{2+}-He^+ implanted W. Figure 4(a) and (b) show the D_2 TDS spectra for only He^+ implanted W and Fe^{2+}-He^+ implanted W with the D fluence of (a) 3.0×10^{22} D^+ m^{-2} or (b) 5.0×10^{24} D^+ m^{-2}. Figure 5 summarizes the damage level dependence on D retention for He^+ implanted W. The D desorption at 400 K for only He^+ implanted W (undamaged W) in Fig. 4(a) was almost three times as high as that for undamaged W in Fig. 2(a). However, the D desorption at 400 K was enhanced by damage introduction. In especially, the D desorption was clearly reduced for 0.3 dpa Fe^{2+}-He^+ implanted W compared to that for 0.3 dpa Fe^{2+} damaged W. It can be said that the accumulation of He near W surface reduces the D diffusion toward the bulk where both of vacancies and voids existed. The D retention for only He^+ implanted W by D plasma exposure was almost the same as that by low D fluence, indicating that the accumulation of He led to the He bubble formation near surface region, whose TEM

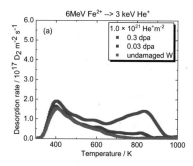 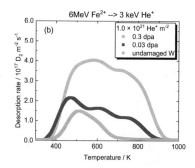

Fig. 4. D_2 TDS spectra for only He^+ implanted W and, 0.03 or 0.3 dpa Fe^{2+}-He^+ implanted W. D fluence of (a) 3.0×10^{22} D^+ m^{-2} or (b) 5.0×10^{24} D^+ m^{-2}

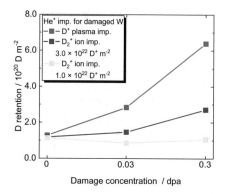

Fig. 5. Damage level dependence on D retention for He^+ implanted W

pictures can be found in Refs. [11, 16], which would reduce D diffusion toward bulk and D trapping by irradiation damages like vacancies and voids. Therefore, for Fe^{2+}-He^+ implanted W in Fig. 4(b), large D desorption was reduced in especially high temperature region around 800 K. Even in the accumulated of radiation-induced damages in W bulk, the formation of He bubbles would work as a D diffusion barrier and enhance the recombination near surface region, leading to the reduction of D retention. In addition, the formation of He-vacancy complex would also reduce D trapping in bulk region. The total D retentions for all the present experiments were summarized in Table 1.

To confirm the proposed mechanism, Hydrogen Isotope Diffusion and Trapping (HIDT) code was applied to simulate the D distribution in damaged W after D plasma exposure [7]. Figure 6 shows HIDT simulation results for W and He^+ implanted W with various damage level by Fe^{2+} implantation. The D trapping energies for the trapping sites for dislocation loops and surface (Peak 1), vacancies (Peak 2) and voids (Peak 3) were summarized in Tables 2 and 3. In the previous study, the existence of He reduces the D trapping energy due to the formation of He-vacancy complex [2, 17, 18]. Therefore, for He^+ implanted W, two trapping energies for Peak 2 were set for

Table 1. Summary of total D retention for all the present experiments

	Undamaged W	0.03 dpa	0.3 dpa
Fe^{2+} implanted W			
D$_2^+$ implantation (1.0 × 10^{22} D m^{-2})	3.29 × 10^{19} D m^{-2}	6.92 × 10^{19} D m^{-2}	1.96 × 10^{20} D m^{-2}
D$_2^+$ implantation (3.0 × 10^{22} D m^{-2})	4.82 × 10^{19} D m^{-2}	1.30 × 10^{20} D m^{-2}	3.87 × 10^{20} D m^{-2}
D plasma exposure (3.0 × 10^{24} D m^{-2})	8.12 × 10^{19} D m^{-2}	3.50 × 10^{20} D m^{-2}	7.41 × 10^{20} D m^{-2}
Fe^{2+}-He$^+$ implanted W			
D$_2^+$ implantation (1.0 × 10^{22} D m^{-2})	1.13 × 10^{20} D m^{-2}	8.61 × 10^{19} D m^{-2}	1.05 × 10^{20} D m^{-2}
D$_2^+$ implantation (3.0 × 10^{22} D m^{-2})	1.18 × 10^{20} D m^{-2}	1.46 × 10^{20} D m^{-2}	2.70 × 10^{20} D m^{-2}
D plasma exposure (3.0 × 10^{24} D m^{-2})	1.28 × 10^{20} D m^{-2}	2.83 × 10^{20} D m^{-2}	6.38 × 10^{20} D m^{-2}

simulation as shown in Fig. 6. It was found that the HIDT simulation represents well the experimental results. For Peak 1 and Peak 2, there was no large change in trapping energy by damage introduction. By the formation of He-vacancy complex, the D trapping energy was reduced around 0.1–0.15 eV, which may be caused by the occupation of He in vacancy. For voids (Peak 3), the accumulation of irradiation damages would stabilize the D trapping and form the most stable trapping sites, leading to the increase of D trapping energy more than 1.4 eV.

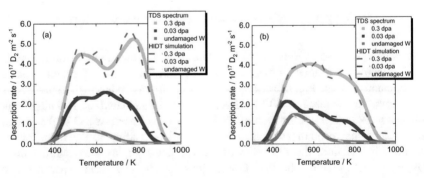

Fig. 6. HIDT simulation results for (a) W and (b) He$^+$ implanted W with various damage concentrations by Fe^{2+} implantation

Table 2. D trapping energies for W after D plasma exposure

Sample	Peak 1	Peak 2	Peak 3
Undamaged W	0.79 eV	1.02 eV	–
0.03 dpa	0.87 eV	1.08 eV	1.30 eV
0.3 dpa	0.90 eV	1.12 eV	1.43 eV

Table 3. D Trapping energies for He^+ implanted W after D plasma exposure

Sample	Peak 1	Peak 2 (He)	Peak 2	Peak 3
Undamaged W	0.80 eV	0.90 eV	1.00 eV	–
0.03 dpa	0.69 eV	0.90 eV	1.05 eV	1.29 eV
0.3 dpa	0.81 eV	1.00 eV	1.15 eV	1.42 eV

4 Conclusions

The irradiation damages and He existence effects on D retention in W were studied with various ion flux and fluence using triple ion implantation system at Shizuoka University and CDPS at Tohoku University. The major D desorption consisted of three stages, around 400, 600 and 800 K, namely the desorption of D trapped by dislocation loops and surface, vacancies and voids, respectively. By the introduction of irradiation damages, the major D desorption stages were shifted towards higher temperature side and the D retentions at 600 and 800 K were increased, indicating that the D was stably trapped by the radiation-induced damages. HIDT simulation showed that the accumulation of irradiation damages stabilized the trapping sites with the trapping energy above 1.4 eV. For He^+ irradiation, D desorption at low temperature was enhanced, due to the formation of dense dislocation loops. By introducing both of radiation-induced damages and He, the D desorption at 800 K was reduced, comparing with that for only Fe^{2+} damaged W. It can be said that the accumulation of He in near surface region with forming He bubbles, which reduces D diffusion toward bulk. In addition, the formation of He-vacancy complex also reduces D trapping in W bulk. It is concluded that both of He bubble formation near surface region and the formation of He-vacancy complex in W bulk contribute on the reduction of D retention in W.

Acknowledgements. This work was performed under the Inter-University Cooperative Research Program of the Institute for Materials Research, Tohoku University (Proposal No. 18M0023) and with the support and under the auspices of the NIFS Collaboration Research program (NIFS17KNWF003). This study was also supported by KAKENHI, 18H03688.

References

1. Ueda, Y., Lee, H.T., Ohno, N., Kajita, S., Kimura, A., Kasada, R., Nagasaka, T., Hatano, Y., Hasegawa, A., Kurishita, H., Oya, Y.: Recent progress of tungsten R&D for fusion application in Japan. Phys. Scr. **T145**, 014029 (2011)
2. Tanabe, T.: Review of hydrogen retention in tungsten. Phys. Scr. **T159**, 014044 (2014)
3. Oya, Y., et al.: Interaction of hydrogen isotopes with radiation damaged tungsten. In: Luca, D., Sirghi, L., Costin, C. (eds.) INTER-ACADEMIA 2017. AISC, vol. 660, pp. 41–49. Springer, Cham (2018). https://doi.org/10.1007/978-3-319-67459-9_6
4. Hatano, Y., Shimada, M., Otsuka, T., et al.: Deuterium trapping at defects created with neutron and ion irradiations in tungsten. Nucl. Fusion **53**, 073006 (2013)
5. Hatano, Y., Shimada, M., Oya, Y., et al.: Retention of hydrogen isotopes in neutron irradiated tungsten. Mater. Trans. **54**, 437–441 (2013)

6. Oya, Y., Shimada, M., Kobayashi, M., et al.: Comparison of deuterium retention for ion-irradiated and neutron-irradiated tungsten. Phys. Scr. **T145**, 14050 (2011)
7. Oya, Y., et al.: Thermal desorption behavior of deuterium for 6 MeV Fe ion irradiated W with various damage concentrations. J. Nucl. Mater. **461**, 336–340 (2015)
8. Fujita, H., et al.: Effect of neutron energy and fluence on deuterium retention behaviour in neutron irradiated tungsten. Phys. Scr. **T167**, 014068 (2016)
9. Fujita, H., et al.: The damage depth profile effect on hydrogen isotope retention behavior in heavy ion irradiated tungsten. Fusion Eng. Des. **125**, 468–472 (2017)
10. Oya, Y., Hatano, Y., Shimada, M., et al.: Recent progress of hydrogen isotope behavior studies for neutron or heavy ion damaged W. Fusion Eng. Des. **113**, 211–215 (2016)
11. Zhou, Q., Azuma, K., Togari, A., et al.: Helium retention behavior in simultaneously He^+–H_2^+ irradiated tungsten. J. Nucl. Mater. **502**, 289–294 (2018)
12. Iwakiri, H., Yasunaga, K., Morishita, K., Yoshida, N.: Microstructure evolution in tungsten during low-energy helium ion irradiation. J. Nucl. Mater. **283**, 1134–1138 (2000)
13. Miyamoto, M., Nishijima, D., Baldwin, M.J.: Microscopic damage of tungsten exposed to deuterium-helium mixture plasma in PISCES and its impacts on retention property. J. Nucl. Mater. **415**, S657–S660 (2011)
14. Ohno, N., Kuwabara, T., Takagi, M., et al.: Development of a compact divertor plasma simulator for plasma-wall interaction studies on neutron-irradiated materials. Plasma Fusion Res **12**, 1405040 (2017)
15. Ziegler, J.F.: The stopping and range of ions in matter. http://www.srim.org/
16. Zhou, Q., Azuma, K., Togari, A., et al.: Deuterium retention behavior in simultaneously He^+-D_2^+ implanted tungsten. Nucl. Mater. Energy (in press)
17. Sato, M., Yuyama, K., Li, X., et al.: Effect of heating temperature on deuterium retention behavior for helium/carbon implanted tungsten. Fusion Sci. Technol. **68**, 531–534 (2015)
18. Alimov, V., Tyburska- Püschel, B., Hatano, Y., et al.: The effect of displacement damage on deuterium retention in ITER-grade tungsten exposed to low-energy, high-flux pure and helium-seeded deuterium plasmas. J. Nucl. Mater. **420**, 370–373 (2012)

Activation of Water by Surface DBD Micro Plasma in Atmospheric Air

Adina Dascalu[1], Alexandra Besleaga[1], Kazuo Shimizu[2],
and Lucel Sirghi[1(✉)]

[1] Faculty of Physics, Iasi Plasma Advanced Research Center (IPARC),
Alexandru Ioan Cuza University of Iasi, Blvd. Carol I nr.
11, 700506 Iasi, Romania
lsirghi@uaic.ro

[2] Organization for Innovation and Social Collaboration, Shizuoka University,
3-5-1, Naka-ku, Johoku, Hamamatsu, Shizuoka 432-8561, Japan

Abstract. Exposure of water to atmospheric discharge plasma in air determines generation of long-living reactive species as hydrogen peroxide H_2O_2, nitrites NO_2^- and nitrates NO_3^-. The water thus treated is called plasma activated water and have applications in medicine and agriculture. In the present work, surface DBD micro plasma working in air at atmospheric pressure is used to treat a small amount of deionized water. The surface DBD plasma was generated on the dielectric surface of a device formed by two silver electrodes deposited on the two sides of a thin glass plate. The device was powered by a high voltage amplifier that applied on the electrodes a sinusoidal waveform voltage at a frequency around 13 kHz and peak to peak amplitude of 4 kV. Time series of current intensity and voltage values were acquired to determine the power injected into the surface DBD micro plasma. Optical emission spectroscopy was used to get information on reactive species generated by micro plasma in the gaseous phase. The reactive species generated in water during the discharge-on and discharge-off periods were investigated by means of UV absorption spectroscopy. The measurements showed that the concentration of these reactive species rise not only during the discharge-on time, but also during discharge-off time for a few tenths of minutes. This observation points towards an optimization scheme of the treatment, which uses discharge on/off steps.

Keywords: Surface DBD · Atmospheric air plasma · Plasma activated water

1 Introduction

Atmospheric pressure plasmas in air generate reactive oxygen and nitrogen species that in contact with water produce secondary reactive species as hydrogen peroxide H_2O_2, nitrites NO_2^-, nitrates NO_3^- and peroxynitrites/peroxynitrous acid $ONOO^-/ONOOH$ [1]. The water treated in this way is called plasma activated water (PAW) and have applications in medicine [2] and agriculture [3]. In this work, PAW is obtained by subjecting a small amount of water to the environment created by a surface dielectric barrier discharge (SDBD) plasma working in a closed volume of air at atmospheric pressure. UV absorption spectroscopy is used to investigate the kinetics of the

© Springer Nature Switzerland AG 2019
G. Laukaitis (Ed.): INTER-ACADEMIA 2018, LNNS 53, pp. 97–104, 2019.
https://doi.org/10.1007/978-3-319-99834-3_13

secondary reactive species generated in PAW during running of SDBD and after the discharge was cut off.

2 Experiment and Results

2.1 Experimental Setup

The SDBD microplasma was generated by a small device (active area of 1.3×3.7 cm^2) consisting in two silver electrodes painted on the two faces of a thin glass lamella (0.2 mm in thickness). The upper (active) electrode had evenly distributed 6×12 array of square holes (2 mm $\times$ 2 mm) while the grounded electrode covered evenly the backside of the glass lamella. The SDBD device was placed in a glass vessel (1 L in volume) side-by-side to a Petri dish (40 mm in diameter) that contained a small amount (5 ml) of the deionized (DI) water to be treated (see Fig. 1). The SDBD was powered by a high voltage amplifier that applied a sinusoidal voltage wave (frequency about 13 kHz) with peak-to-peak value of about 4000 V. The voltage and intensity of discharge current signals were collected by specialized voltage and intensity probes and monitored by a digital oscilloscope (Tektroniks, USA).

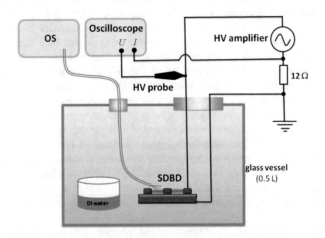

Fig. 1. Sketch of the experimental setup used to treat a small amount of deionized (DI) water by SDBD plasma. The SDBD current voltage and intensity waveforms are monitored by an oscilloscope. The light emitted by SDBD plasma is collected by an optical fiber and fed into an optical spectrometer (OS)

2.2 Electric Characterization of SDBD

Time variations of voltage and current intensity during an oscillation period are shown in Fig. 2. In absence of plasma (voltage below gas breakdown value) the device behaves as a capacitor with power loss in the glass dielectric. By increasing the voltage applied on the device, the SDBD is ignited and generates microplasma at the edges of the structured electrodes. The discharge current intensity shows mainly a capacitive and resistive current with a sinusoidal time variation with the amplitude of about 32 mA and a phase advance of about 83° with respect to the phase of the voltage waveform. The capacitive and resistive loss current intensities determined values of 196 pF and 557 kΩ for the device capacitance of electric resistance, respectively. Apart of the low-frequency sinusoidal time variations, the current intensity shows much faster variations with spikes typical for multi-filamentary discharges. The spike amplitude and number are larger for the positive discharge, when the voltage applied on hot electrode is positive with respect to the grounded electrode, than for the negative discharge.

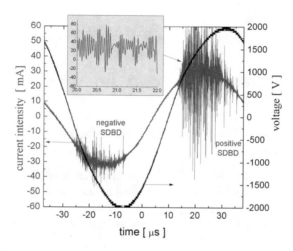

Fig. 2. Current intensity and voltage waveforms recorded during an oscillation period of the SDBD. The inset gives details on the current intensity time variations during 2 μs

Typical values for the amplitude and lifetime of spikes are around 20 mA and 20 ns, respectively. The high frequency oscillations fallowing each spike in the discharge current intensity are due to the inductance of electric circuit formed by the high-voltage amplifier, connection wires and plasma. Because of these oscillations is difficult to identify spikes corresponding to individual plasma filaments formed in SDBD.

The energy dissipated by SDBD microplasma during one oscillation period and the average discharge power are determined from charge-voltage Lissajous diagram (see Fig. 3) to about 200 μJ and 2.5 W, respectively. The diagram considers only the electric charge transported by plasma by extracting from the total current intensity through the SDBD device the capacitive and resistive current intensities through the dielectric [4]. Thus, the diagram shows that the electric charge building up on the dielectric surface during negative or positive discharges was around 50 nC, the charge flowing through the filamentary microplasma during each discharge being around 100 nC.

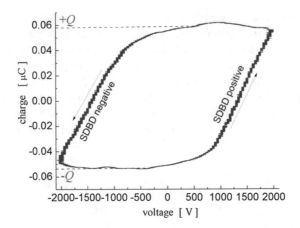

Fig. 3. Charge-voltage diagram for SDBD obtained after the capacitive and resistive current intensities were extracted from the total current intensity through the SDBD device

2.3   Optical Emission Spectroscopy Characterization of SDBD Microplasma

The light emission spectra of SDBD microplasma were acquired in the range 190–650 nm with a resolution of 0.2 nm by an optical spectrometer (HR2000+ from Ocean Optics Inc. USA). The optical emission spectrum of the SDBD plasma in the range 280–440 nm is shown in Fig. 4. The most intense emission bands belong to the second positive system emission of N_2 molecules ($C^3\Pi_u$-$B^3\Pi_g$), and OH radicals ($A^2\Sigma$-$X^2\Pi$, 0–0) with the band head at 306.4 nm. The exact contribution of OH ($A^2\Sigma$-$X^2\Pi$, 0–0) transitions to the light emission spectrum of SDBD plasma is difficult to estimate because superposition of this band with the vibrational transition band of N_2 ($C^3\Pi_u$-$B^3\Pi_g$, 0+1) [5]. The emission intensity corresponding to N_2 vibrational ($C^3\Pi_u$-$B^3\Pi_g$)

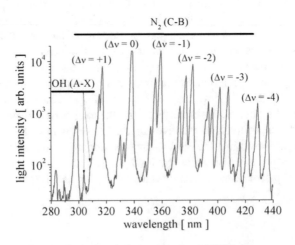

Fig. 4. Optical emission spectrum of SDBD plasma

transitions can be used to determine the vibrational temperature of N_2 in SDBD microplasma [5], which in the present experiment was estimated to about 2640 K.

2.4 UV Absorption Spectroscopy Measurements of blank;Plasma Activated Water

Presence of reactive species in PAW has been determined by measurements of optical absorbance in the UV region (190–450 nm) by a UV-VIS spectrophotometer (Evolution 300 from Thermo Scientific). The PAW and DI water were loaded in quartz cuvettes and mounted to spectrophotometer apparatus as sample and reference, respectively. Figure 5 is showing the absolute UV absorption spectra of DI water and PAW (taken with reference to air) and the UV absorption spectrum of PAW taken with reference to DI water. The UV absorption peaks of the radical species present in the PAW determine a convoluted absorption peak. Deconvolution of the spectra to a sum of individual Gaussian absorption peaks can be used to determine the contributions of H_2O_2, NO_2 and NO_3 species [6]. However, because it is difficult to determine by this method the unique contribution of each kind of radical molecules, we used the area of the whole absorption spectrum as a measure of concentration of radical molecules in PAW.

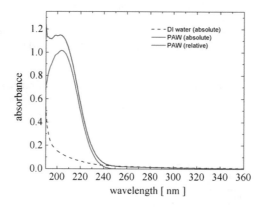

Fig. 5. UV absorption spectra of DI water PAW with respect of air (absolute) and the UV absorption spectrum of PAW with respect of DI water (relative)

Diffusion of reactive species produced by SDBD plasma in air and their interaction with water is a relatively slow process. This is indicated by the fact that the concentration of reactive species in the water kept in the post-discharge medium continues to rise after the SDBD was cut off. Figure 6 shows the dependence of the UV absorption area of PAW kept for various time intervals in the post-discharge medium generated by SDBD plasma in 1 and 5 min, respectively. The plots indicate that in about 20 min after the discharge was cut off, the reactive species in PAW rise about 3 times for the discharge time of 1 min and about 2 times for the discharge time of 5 min. After 20 min, the rise of concentration is negligible small. These results indicate that the kinetics of reactive species produced by plasma in air and water can be used to optimize the treatment. Thus, using a treatment scheme based on discharge on/off cycles can improve the efficiency of treatment in generation of reactive species in PAW. A result of such a treatment scheme is given in Fig. 7, which shows the evolution of total area of the UV absorption spectrum of PAW after various numbers of cycles consisting in 1 min discharge-on and 5 min discharge-off.

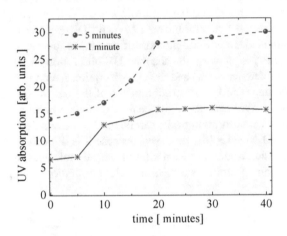

Fig. 6. Variation of UV absorption peak area corresponding to plasma generated radicals in PAW in the post discharge medium for discharge-time values of 5 min and 1 min

The plot in Fig. 7 shows that initially the concentration of reactive species in PAW increases linearly with the number of discharge on/off cycles. However, after 15 cycles the concentration rise speeds up in a nonlinear manner. This is happening because

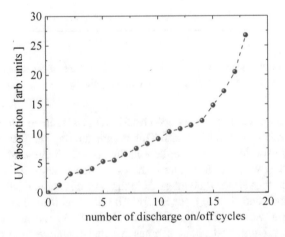

Fig. 7. Variation of area of UV absorption spectra corresponding to plasma generated radicals in PAW during water treatment in steps of 1 min discharge on/5 min discharge off

accumulation of reactive species in the gas phase. Further experiments on variation of radical concentrations in PAW obtained by the treatment scheme based on discharge on/off cycles are necessary in order to establish the optimum ratio between discharge-on and discharge-off times.

3 Conclusion

Surface DBD plasma working in a closed volume of air at atmospheric pressure was used to activate water. The surface DBD plasma was generated at the surface of the glass dielectric of a small surface DBD device by applying a sinusoidal voltage (about 4 kV in peak-to-peak amplitude) to the silver electrodes painted on the glass sides. Time series of voltage and discharge current intensity values indicated a multi filamentary surface DBD with about 4 W in power and about 60 nC in electric charge of build up at the dielectric surface during each discharge. The optical emission spectra of the surface DBD plasma showed mainly emission bands corresponding to N_2 molecule transitions ($C^3\Pi_u$-$B^3\Pi_g$). The intensities of these emission bands were used to determine the vibrational temperature of N_2 in surface DBD microplasma, which was around 2600 K. Occurrence of a weak emission band with the band head at 306.4 nm in the optical emission spectra indicated presence of OH radicals ($A^2\Sigma$-$X^2\Pi$, 0–0) in the gas phase of surface microplasma.

The reactive oxygen and nitrogen species generated by plasma in gas phase diffuse and react with water to generate secondary reactive species as H_2O_2, NH_2 and NH_3. Presence of these species in plasma activated water is probed by UV absorption spectrometry measurements. The UV absorption spectra of the plasma activated water showed absorption patterns corresponding to convolution of individual peaks attributed to H_2O_2, NH_2 and NH_3 radical species. Because it is difficult to determine the unique contribution of each kind of radical molecules to the total UV absorption spectra, the area of the whole absorption spectrum was used as a measure of concentration of radical molecules in PAW. These measurements indicated that the concentration of radical species continues to rise for few tenths of minutes in the water kept in the post-discharge medium (after the discharge was cut off). Therefore, an optimization of the treatment in terms of treatment time and energy consumption is possible when it is adopted a treatment scheme based on cycles comprising discharge-on and discharge-off periods.

References

1. Park, D.P., Devis, K., Gilani, S., Alonzo, C.A., Drexel, A.J.: Reactive nitrogen species produced in water by non-equilibrium plasma increase plant growth rate and nutritional yield. Curr. Appl. Phys. **13**, S19–S29 (2013)
2. Fridman, G., Friedman, G., Gutsol, A., Shekhter, A.B., Vasilets, V.N., Fridman, A.: Applied plasma medicine. Plasma Process. Polym. **5**, 503–533 (2008)

3. Sivachandirana, L., Khacef, A.: Enhanced seed germination and plant growth by atmospheric pressure cold air plasma: combined effect of seed and water treatment. RSC Adv. **7**, 1822–1832 (2017)
4. Kostov, K.G., Honda, R.Y., Alves, L.M.S., Kayama, M.E.: Characteristics of dielectric barrier discharge reactor for material treatment. Braz. J. Phys. **39**(2), 322–325 (2009)
5. Liu, F., Wang, W., Wang, S., Zheng, W., Wang, Y.: Diagnosis of OH radical by optical emission spectroscopy in a wire-plate bi-directional pulsed corona discharge. J. Electrostat. **65**, 445–451 (2007)
6. Oh, J.-S., Ito, S., Furuta, H., Hatta, A.: Time-resolved in situ UV absorption spectroscopic studies for detection of reactive oxygen and nitrogen species (RONS) in plasma activated water. In: 22nd International Symposium on Plasma Chemistry, Antwerp, Belgium (2015)
7. Lukes, P., Dolezalova, E., Sisrova, I., Clupek, M.: Aqueous-phase chemistry and bactericidal effects from an air discharge plasma in contact with water: evidence for the formation of peroxynitrite through a pseudo-second-order post-discharge reaction of H_2O_2 and HNO_2. Plasma Sources Sci. Technol. **23**, 015019 (2014)
8. Dascalu, A., Demeter, A., Samoila, F., Anita, V., Shimizu, K., Sirghi, L.: Surface dielectric barrier discharge in closed-volume air. Plasma Med. **4**, 395–406 (2017)

Design and Creation of Metal-Polymer Absorbing Metamaterials Using the Vacuum-Plasma Technologies

Igor Semchenko[1]($\boxtimes$), Sergei Khakhomov[1] (ID), Andrey Samofalov[1],
Ihar Faniayeu[1], Dzmitry Slepiankou[1], Vitaliy Solodukha[2],
Alyaksandr Pyatlitski[2], Natalya Kovalchuk[2], Andrey Goncharenko[3],
and George Sinitsyn[3]

[1] Francisk Skorina Gomel State University, Sovyetskaya Str. 104,
246019 Gomel, Belarus
isemchenko@gsu.by
[2] JSC "INTEGRAL", Korjenevsky Str. 12, 220108 Minsk, Belarus
[3] B.I. Stepanov Institute of Physics, Nezavisimosty Av. 68,
220072 Minsk, Belarus

Abstract. The objective of the paper is to create on the basis of omega-shaped bianisotropic elements new absorbing metamaterials and coatings that do not have a reflecting base and are "invisible" on the irradiated side.

Keywords: Metamaterials · Omega-elements · Reflection

1 Introduction

The unusual properties of metamaterials were experimentally studied mainly in the MHz and GHz ranges, where the resonant metamaterial elements are of millimeter and centimeter sizes [1–7]. It is easy to form three-dimensional (3D) elements and arrange them in the form of 3D arrays. Currently, there is a clear trend towards the development and investigation of THz metamaterials, since that THz technique is rapidly developing [8–11]. However, the realization of absorbing metamaterials based on 3D resonators for THz is quite difficult task since it requires specialized expensive fabrication technologies. In this paper we are aimed to create a new absorbing metamaterial and coatings based on 3D bianisotropic striped omega-resonators for microwave range that do not have a reflecting base and are "invisible" on the irradiated side. Designed low-reflection functional metamaterials and metasurfaces have been previous studied in our works [12–18].

To obtain a matched metamaterial impedance, all resonators of a very large array should be tuned very accurately. Among the widely used techniques, only the conventional planar technology (which allows us to form planar elements and their layers) can provide the required sizes and accuracy. In principle, it is impossible to specify the properties of such a metamaterial consisting of planar elements in all three dimensions. Moreover, in most experiments the researchers have to be restricted to one layer of elements (i.e., metamaterial monolayer) because of the restrictions of the planar

G. Laukaitis (Ed.): INTER-ACADEMIA 2018, LNNS 53, pp. 105–112, 2019.
https://doi.org/10.1007/978-3-319-99834-3_14

technology, which hinders the study of volumetric electromagnetic properties. At the same time, volumetric metamaterials with specified three-dimensional electromagnetic properties are necessary for almost all current metamaterial applications.

2 Simulation

Using a standard technology for the fabrication of planar architectures it is possible to create a volumetric metamaterial using strip-line elements with an arbitrary pattern. Figure 1 shows the basic idea how to design a volumetric metamaterials using a striped bianisotropic element. Design of this structure is a flat element in the form of a Greek letter "omega" of a rectangular shape arranged on the transparent dielectric (it does not influence to the microwave radiation) as shown in Fig. 1(a). We propose the design of a metamaterial based on such flat resonators that are in a periodic array in a staggered order. The orientation of the omega-element loops alternates, and they are directed up and down in the metamaterial as shown in Fig. 1(b).

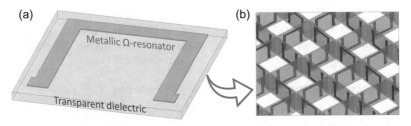

Fig. 1. (a) Omega element of rectangular shape. (b) The schema of construction of the 3D sample from planar stripes. On the adjacent strips, the orientation of the omega-element loops alternates, they are directed up and down

We have proposed a flat structure for creating an ideal absorber in the microwave range. The element of such a structure is a flat element in the form of a Greek letter "omega" of rectangular shape (Fig. 2(a)).

The characteristics of the electromagnetic radiation scattered on such an element depend on the ratio of the geometric parameters a, b, d and wavelength λ. Let us consider the case when the linear geometric dimensions of the omega-element are much smaller than the length of the incident wave, which makes it possible to apply the dipole approximation of the radiation theory. For this case, we find the electric dipole and magnetic moments induced in the omega element under the action of the incident electromagnetic wave. Simultaneous occurrence in the element of both the electric dipole and magnetic moments, interconnected and induced by the external field, is the main condition for displaying the interesting properties of such a structure. We first consider the case b = d, that is, the shape of the element is close to a square. In this case, the area bounded by the element is relatively large, and the resulting magnetic moment has a significant value. In this case, we assume that the thickness of the strip h is negligibly small in comparison with the lengths of the sides b and d.

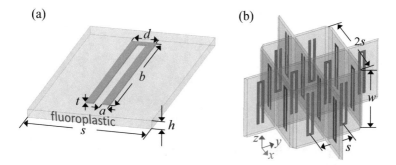

Fig. 2. (a) A single strip Ω-resonator on the fluoroplastic substrate with the following structural parameters: $a = 1$ mm, $b = 22$ mm, $d = 3$ mm, $t = 17.5$ μm, $h = 1$ mm, and $s = 10$ mm. (b) The unit cell of absorber based on proposed resonators with indicated parameters ($w = 30$ mm)

The dipole electric moment inducing in an omega element is equal to

$$p = ql = q(2a + b),\tag{1}$$

here q = −e is the electron charge.

The magnetic moment of an element induced by an electric field is

$$m = IS = \frac{dq}{dt}b^2 = j\omega b^2 q.\tag{2}$$

To calculate the parameters of the omega element, let us use the expression for the components of the electric dipole moment and magnetic moment, the condition of the main frequency resonance (the length of the conductor L is equal to half of the length of the incident wave λ)

$$L = \frac{\lambda}{2},\tag{3}$$

and the universal relation for the components of the electric dipole and magnetic moments obtained in the articles [19–22]

$$|p| = \frac{|m|}{c}.\tag{4}$$

Substituting the expressions (1) and (2) into (4), we obtain

$$|q|(2a + b) = \frac{1}{c}|j|\omega b^2|q|, \quad 2a + b = \frac{1}{c}\omega b^2.\tag{5}$$

Taking into account the condition of the main frequency resonance we obtain

$$2a + 3b = \frac{\lambda}{2} = \frac{\pi c}{\omega}. \tag{6}$$

As a result, we obtain a system of equations

$$\begin{cases} 2a + b = \frac{1}{c}\omega b^2 \\ 2a + 3b = \frac{\lambda}{2} = \frac{\pi c}{\omega} \end{cases} \tag{7}$$

Solving the system of Eq. (7) we can obtain:

$$b = \frac{\lambda}{2\pi}\left(\sqrt{1+\pi} - 1\right). \tag{8}$$

$$a = \frac{\lambda}{4}\left(\pi + 3 - 3\sqrt{\pi+1}\right). \tag{9}$$

Figure 2 shows the design of striped omega-resonators (a) with optimized structural parameters that builds a low-reflection volumetric metamaterial (b). These optimized resonators have a non-square, but rectangular shape. The substrate is a low permittivity of 2.1 fluoroplastic which serves as a mechanical support of the resonators. The material of resonators is titanium since it has enough dissipative loss at the excitation by external field. The numerical simulation based on Finite Element Method (FEM) was used to find out the reflectance (R), the transmittance (T), and absorbance (A) of a proposed metamaterial design. First, we are tailoring the structural parameters in accordance with analytical descriptions above to achieve the total absorption while the metamaterial should not have a reflection in the broad microwave range.

Also, the possibility of changing the direction of the shoulders (inside or outside) of the omega elements and the square of the conductor section is realized.

After constructing an array model consisting of omega elements, in accordance with the calculated parameters, boundary conditions and parameters of the incident electromagnetic wave were set. When solving the problem, a plane incident wave was used. In our case we have linearly polarized waves with mutually perpendicular polarizations. The field vectors for the TM and TE modes lie in the XOY plane, and the wave propagates along the OZ axis Fig. 2(b).

Figure 3 shows the reflection, transmission, and absorption spectra of the electromagnetic wave for TE (a) and TM (b) polarizations at normal incidence. As seen from simulated R, T, A spectra, absorption peak reaches a maximum value of 0.75 at the resonant frequency of 2.74 GHz, while reflection does not exceed of 0.067 at the operational frequency range from 2 to 4 GHz. The simulated spectra are identical for the TE and TM polarizations, therefore, the proposed metamaterial does not depends on the polarization of incident waves. Moreover, metamaterial consisting of omega-resonators exhibits near perfect reflectionless behavior and transparency away from the resonant band.

Next, we estimate the calculated absorption spectra for TE and TM polarizations as a function of titanium thickness (t) dependence of omega-based metamaterial that are illustrated in Fig. 4(a, b). As one can see, absorbance peaks reach maximum values of

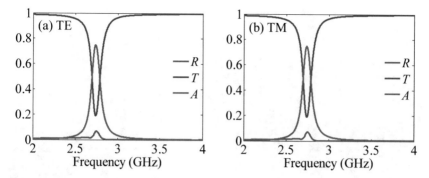

Fig. 3. The simulated spectra of reflectance, transmittance, and absorbance for TE (a) and TM (b) polarizations at normal incidence

0.82 in the resonance 2.77 GHz at the thickness of 70 μm. The additional peaks can be seen in the spectra which is an undesirable effect for this absorber. Nevertheless, this does not affect the performance of the absorber based on omega resonators in the microwave range (Fig. 5).

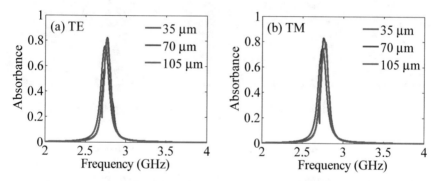

Fig. 4. Reflection, transmission, and absorption spectra of absorbing metamaterial for TE (a) and TM (b) polarizations at different thickness (t) of Ω-resonators

Thus, we have outlined the basic mechanisms responsible for the occurrence of absorption resonances in striped omega-based metamaterial, as well as their main design principles with volumetric metamaterials.

3 Manufacturing

The mask was made in a stainless steel plate 08X18H10 using a solid-state YAG laser (wavelength $\lambda = 1.064$ μm, power P = 50 W, maximum laser pulse energy 790 mJ, divergence 0.8 mrad, frequency 50 Hz, and pulse duration 1 ms). As a vacuum-plasma technique for depositing copper on a substrate, the magnetron sputtering method was chosen [23].

Fig. 5. The process of mask manufacturing, vacuum chamber for magnetron sputtering

Omega elements of rectangular shape were made of copper applied to fluoroplastic (Fig. 6).

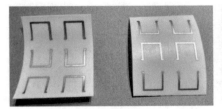
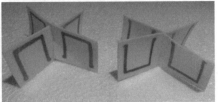

Fig. 6. Omega elements of rectangular shape made of copper applied to fluoroplastic

The next stage of the research is to study artificial experimental samples in an anechoic chamber at Francisk Skorina Gomel State University in the microwave range and to compare the simulation with the experiment. Further, using the principle of electromagnetic scaling, a structure for creating an ideal absorber in the THz range will be proposed.

4   Conclusions

Using our approach and numerical simulation, we have found the design of absorbing volumetric metamaterial based on 3D striped omega-resonators for microwave range. The numerically simulated results have shown that absorber reaches peak value over 82% in vicinity of the resonant frequency band of 2.77 GHz and exhibits low reflection in the operational frequencies of 2–4 GHz, respectively. The findings can be used for theoretical and experimental studies of artificial composite absorbing media with inclusions of various shapes, in the development of microwave absorbers with improved properties.

Acknowledgements. This study was supported by the Belarusian Republican Foundation for Fundamental Research, projects F18KI-027, F18KI-028.

References

1. Landy, N.I., Sajuyigbe, S., Mock, J.J., Smith, D.R., Padilla, W.J.: Perfect metamaterial absorber. Phys. Rev. Lett. **100**, 207402 (2008)
2. Gansel, J.K., Thiel, M., Rill, M.S., Decker, M., Bade, K., Saile, V., Freymann, G., Linden, S., Wegener, M.: Gold helix photonic metamaterial as broadband circular polarizer. Science **325**, 1513 (2009)
3. Zhao, Y., Belkin, M.A., Alù, A.: Twisted optical metamaterials for planarized ultrathin broadband circular polarizers. Nat. Commun. **3**, 870 (2012)
4. Ra'di, Y., Tretyakov, S.A.: Balanced and optimal bianisotropic particles: maximizing power extracted from electromagnetic fields. New J. Phys. **15**, 053008 (2013)
5. Niemi, T., Karilainen, A., Tretyakov, S.: Synthesis of polarization transformers. IEEE Trans. Antennas Propag. **61**(6), 3102–3111 (2013)
6. Ra'di, Y., Asadchy, V.S., Tretyakov, S.A.: Tailoring reflections from thin composite metamirrors. IEEE Trans. Antennas Propag. **62**(7), 3749–3760 (2014)
7. Ra'di, Y., Asadchy, V.S., Tretyakov, S.A.: One-way transparent sheets. Phys. Rev. B **89**, 075109 (2014)
8. Semchenko, I.V., Khakhomov, S.A., Samofalov, A.L., Podalov, M.A., Solodukha, V.A., Pyatlitski, A.N., Kovalchuk, N.S.: Omega-structured substrate-supported metamaterial for the transformation of wave polarization in THz frequency range. In: Luca, D., Sirghi, L., Costin, C. (eds.) Recent Advances in Technology Research and Education, INTER-ACADEMIA 2017. Advances in Intelligent Systems and Computing, vol. 660, pp. 72–80. Springer (2018)
9. Semchenko, I.V., Khakhomov, S.A., Samofalov, A.L., Podalov, M.A., Songsong, Q.: The effective optimal parameters of metamaterial on the base of omega-elements. In: Jablonski, R., Szewczyk, R. (eds.) Recent Global Research and Education: Technological Challenges, INTER-ACADEMIA 2016. Advances in Intelligent Systems and Computing, vol. 519, pp. 3–9. Springer (2017)
10. Semchenko, I.V., Khakhomov, S.A., Asadchy, V.S., Golod, S.V., Naumova, E.V., Prinz, V. Ya., Goncharenko, A.M., Sinitsyn, G.V., Lyakhnovich, A.V., Malevich, V.L.: Investigation of electromagnetic properties of a high absorptive, weakly reflective metamaterial-substrate system with compensated chirality. J. Appl. Phys. **121**(1), 015108 (2017)
11. Semchenko, I.V., Khakhomov, S.A., Naumova, E.V., Prinz, V.Ya., Golod, S.V., Kubarev, V.V.: Study of the properties of artificial anisotropic structures with high chirality. Crystallogr. Rep. **56**(3), 366–373 (2011)
12. Faniayeu, I.A., Khakhomov, S.A., Semchenko, I.V., Mizeikis, V.: Highly transparent twist polarizer metasurface. Appl. Phys. Lett. **111**, 111108 (2017)
13. Asadchy, V.S., Faniayeu, I.A., Ra'di, Y., Khakhomov, S.A., Semchenko, I.V., Tretyakov, S. A.: Broadband reflectionless metasheets: frequency-selective transmission and perfect absorption. Phys. Rev. X **5**, 031005 (2015)
14. Semchenko, I.V., Balmakou, A.P., Khakhomov, S.A., Tretyakov, S.A.: Stored and absorbed energy of fields in lossy chiral single-component metamaterials. Phys. Rev. B **97**, 014432 (2018)

15. Asadchy, V.S., Faniayeu, I.A., Semchenko, I.V., Khakhomov, S.A., Ra'di, Y.: Optimal arrangement of smooth helices in uniaxial 2D-arrays. In: Metamaterials' 2013 Proceedings, pp. 244–246, Bordeaux, France (2013)
16. Semchenko, I.V., Khakhomov, S.A., Tretyakov, S.A., Sihvola, A.H.: Electromagnetic waves in artificial chiral structures with dielectric and magnetic properties. Electromagnetics **21**(5), 401–414 (2001)
17. Semchenko, I.V., Khakhomov, S.A.: Artificial uniaxial bi-anisotropic media at oblique incidence of electromagnetic waves. Electromagnetics **22**(1), 71–84 (2002)
18. Balmakou, A., Podalov, M., Khakhomov, S., Stavenga, D., Semchenko, I.: Ground-plane-less bidirectional terahertz absorber based on omega resonators. Opt. Lett. **40**(9), 2084–2087 (2015)
19. Semchenko, I.V., Khakhomov, S.A., Podalov, M.A., Tretyakov, S.A.: Radiation of circularly polarized microwaves by a plane periodic structure of Ω elements. J. Commun. Technol. Electron. **52**(9), 1002–1005 (2007)
20. Semchenko, I.V., Khakhomov, S.A., Samofalov, A.L.: Optimal helix shape: equality of dielectric, magnetic, and chiral susceptibilities. Russ. Phys. J. **52**, 472–479 (2009)
21. Semchenko, I.V., Khakhomov, S.A., Samofalov, A.L.: Transformation of the polarization of electromagnetic waves by helical radiators. J. Commun. Technol. Electron. **52**(8), 850–855 (2007)
22. Semchenko, I.V., Khakhomov, S.A., Tretyakov, S.A.: Chiral metamaterial with unit negative refraction index. Eur. Phys. J. Appl. Phys. **46**(3), 32607 (2009)
23. Danilin, B.S., Syrchin, V.K.: Magnetron sputtering systems. Radio and communication, Moscow (1982)

Modification of Optical Properties of Amorphous Metallic Mirrors Due to Impact of Deuterium Plasma

Inna Lyashenko[1], Vladimir Konovalov[2], Vladimir Lopatka[1],
Leonid Poperenko[1], Ivan Ryzhkov[2], Vladimir Voitsenya[2],
and Iryna Yurgelevych[1(✉)]

[1] Faculty of Physics, Taras Shevchenko National University of Kyiv,
Academician Glushkov Avenue 2, Building 1, Kyiv 03680, Ukraine
vladira_19@ukr.net
[2] Institute of Plasma Physics of NSC KIPT,
1, Akademicheskaya St., Kharkov 61108, Ukraine

Abstract. Optical properties of zirconium-based amorphous alloys that were subjected to deuterium plasma treatment have been investigated by the multiple-angle-of-incidence single-wavelength and spectral ellipsometry in a spectral range of 0.5–3.5 eV. After bombardment by ions of deuterium plasma of the samples the increase in the intensity of absorption within the spectra of optical conductivity in the indicated spectral range was found. Such behavior of the optical properties can be explained by deuterium plasma ions sputtering of a subsurface layer, and by modification of the electronic properties of a near-surface layer due to deuterium implantation. The optical anisotropy of the samples both in the initial and after treatment in the deuterium plasma was not observed being inherent for amorphous surface layer structures. It has been found that the treatment of mirror-like surfaces by the deuterium plasma increases the roughness of the surface of $Zr_{57}Cu_{15.4}Al_{10}Ni_{12.6}Nb_5$ alloy mirror sample investigated.

Keywords: Amorphous metal alloys · Deuterium plasma · Ellipsometry

1 Introduction

Amorphous metallic alloys due to unique combination of their physical, chemical and mechanical properties are promising for practical use in the different fields of science and industry [1]. Previously, their use was limited due to the inability to obtain bulk materials by existing methods of production of amorphous metallic alloys based on rapid quenching from the melt at the rates of the order 10^6 K/s.

During the last several decades intensive progress was made and the technology of production of the multicomponent amorphous alloys with significantly reduced critical cooling rates to prevent crystallization have been developed [2]. These new alloys, identified as bulk metallic glasses (BMGs) could be produced at thicknesses more than 1 mm [3]. A number of applications have been suggested for these materials, namely casing in cellular phones, casing in electro-magnetic instruments, connecting part for

© Springer Nature Switzerland AG 2019
G. Laukaitis (Ed.): INTER-ACADEMIA 2018, LNNS 53, pp. 113–120, 2019.
https://doi.org/10.1007/978-3-319-99834-3_15

optical fibers, soft magnetic choke coils, biomedical instruments, various shapes of optical mirrors, etc. [2]. The perspective of application of the Zr-based bulk metallic glasses as first mirrors in optical schemes of plasma diagnostics in fusion reactor has been substantiated in the work [4]. The effects of bombardment by ions of deuterium plasma on a reflectance of mirror samples were studied. It was concluded that optical characteristics of BMGs mirrors are more stable and resistive under ion bombardment in comparison with crystallized metal ones subjected to such action.

In order to successfully apply the bulk metallic glasses, it is necessary to characterize their optical properties and atomic and electronic structure by different experimental methods. In this work, the ellipsometry as effective method for determining optical constants of amorphous metallic alloys was used [5] for investigation of optical properties of the Zr-based BMG mirror samples under bombardment by ions of deuterium plasma.

2 Experimental

The optical properties of $Zr_{57}Cu_{15.4}Al_{10}Ni_{12.6}Nb_5$ (alloy-1) and $Zr_{48}Cu_{36}Al_8Ag_8$ (alloy-2) mirrors subjected to deuterium plasma action have been studied. Previously, the mirrors were polished to high optical quality and purified with low-energy ions of deuterium plasma from organic films that could have formed on the surface of the mirrors after washing them in alcohol after polishing and long time saving in air. The detail description of the experimental procedure of performing the ion bombardment of these mirrors can be found in [4]. The ion energy of D plasma was fixed at 100 eV for all samples, excluding one of $Zr_{48}Cu_{36}Al_8Ag_8$ samples (AM-J_p(0-1) sample) where the ion energy was increased from 30 to 600 eV with every next exposure (total ion fluence was 15×10^{24} ions/m^2). The more detail specification of the samples is presented in the Table 1.

Two sides (conditionally named as the face surface and back surfaces) of the unaffected sample (AMGN2) and sample treated by deuterium plasma (AMGN1) of $Zr_{57}Cu_{15.4}Al_{10}Ni_{12.6}Nb_5$ alloy and one side of the $Zr_{48}Cu_{36}Al_8Ag_8$ alloy samples such as AM-J_p(0-1) (within different surface areas 1 and 2), AM-J_pN2 and AM-J_pN4 were investigated by multiple-angle-of-incidence single-wavelength and spectral ellipsometry and atomic force microscopy (AFM). The ellipsometric measurements of the phase shift Δ between the p- and s-components of the polarization vector and the azimuth Ψ of the restored linear polarization for these mirrors were obtained as dependences on the light incidence angle φ using a laser ellipsometer LEF-3M-1 with $\lambda = 632.8$ nm. On the basis of the obtained data the principal angles of incidence φ_p ($\cos\Delta = 0$) and the angular position of the minimal value of tgΨ_{min} for the investigated samples were calculated. For all mirror samples the values φ_p and tgΨ_{min} were obtained for two mutually perpendicular directions in own plane of the sample with an aim to determine the optical anisotropy within surface layer of the mirror. The inaccuracies of the determined data were as follows: $\delta\varphi_p = \pm0.05°$ and δtg$\Psi_{min} = \pm0.005$.

Spectral ellipsometric measurements of such ellipsometric parameters of the mirrors were performed on the automated spectral ellipsometer at a given angle of light incidence of 73° in the range of wavelengths 500–740 nm (photon energy $E = 0.5$–

3.5 eV). The dependences of optical parameters, namely the refraction n and absorption k indices, a real ε_1 and an imaginary ε_2 parts of the dielectric function as well as optical conductivity σ on the photon energy E were obtained. The latter are connected with ellipsometric parameters Δ and Ψ in the model of a filmless semi-infinite reflective medium by relations [6]:

$$n^2 - k^2 = \sin^2\phi\left[1 + \mathrm{tg}^2\phi\,\frac{\cos^2 2\Psi - \sin^2 2\Psi \sin^2\Delta}{(1 + \sin 2\Psi \cos\Delta)^2}\right],\tag{1}$$

$$2nk = \sin^2\phi\,\mathrm{tg}^2\phi\,\frac{\sin 4\Psi \sin\Delta}{(1 + \sin 2\Psi \cos\Delta)^2},\tag{2}$$

$$\varepsilon_1 = n^2 - k^2, \quad \varepsilon_2 = 2nk,\tag{3}$$

$$\sigma = nk\nu\tag{4}$$

where ν is a linear frequency of light.

The AFM measurements were carried out using microscope INTEGRA NT-MDT.

3   Results and Discussion

The typical curves of $\cos\Delta(\varphi)$ and $\mathrm{tg}\Psi(\varphi)$ for one side of alloy-1 mirror samples (unaffected and treated by deuterium plasma) are presented in Fig. 1. The behavior of $\cos\Delta(\varphi)$ and $\mathrm{tg}\Psi(\varphi)$ curves for another side of samples is similar.

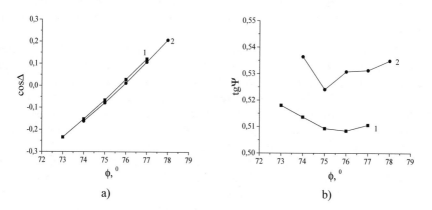

Fig. 1. Dependences of $\cos\Delta(\varphi)$ (a) and $\mathrm{tg}\Psi(\varphi)$ (b) for face surface of 1 – AMGN1 and 2 – AMGN2 mirror samples

One can see that curves of $\cos\Delta(\varphi)$ and $\mathrm{tg}\Psi(\varphi)$ for AMGN1 sample treated by deuterium plasma are shifted to the left and down respectively. The obtained values of

Table 1. The principal angles of incidence φ_p, minimal values of $tg\Psi_{min}$, refraction n and absorption k indices for $Zr_{57}Cu_{15.4}Al_{10}Ni_{12.6}Nb_5$ and $Zr_{48}Cu_{36}Al_8Ag_8$ alloys samples

Sample		φ_p, °	$tg\Psi_{min}$	n	k
AMGN1, 20 × 22 mm²	Face surface	75.69; 75.76	0.5084; 0.5068	2.24	3.07
	Back surface	75.95; 75.84	0.5088; 0.5106	2.28	3.13
AMGN2, 20 × 22 mm²	Face surface	75.85; 75.85	0.5240; 0.5250	2.19	3.16
	Back surface	76.09; 76.13	0.5269; 0.5261	2.21	3.23
AM-J_p(0-1) – lentils, Ø 20 mm	Surface area 1	74.67	0.5658	1.80	3.00
	Surface area 2	74.31	0.5465	1.84	2.87
AM-J_pN2, convex, Ø 19.9 mm, R = 75 cm		73.06; 73.05	0.5764; 0.5692	1.57	2.68
AM-J_pN4, concave, Ø 20 mm, R = 88 cm		73.69; 73.62	0.5627; 0.5646	1.64	2.85

other ellipsometric parameters such as φ_p and $tg\Psi_{min}$, refraction n and absorption k indices are shown in the Table 1.

Comparing values φ_p and $tg\Psi_{min}$ of samples AMGN1 and AMGN2, it was observed that for the former sample these parameters are lower than for both sides of AMGN2 sample, which was not exposed in plasma. One of the reasons can be some increase of the AMGN1 sample surface roughness after treatment by deuterium plasma. This fact follows from our previous results [5] where was found an interconnection between behavior of absolute values of $tg\Psi_{min}$ and roughness height for the bulk sample of nickel, namely the $tg\Psi_{min}$ decreases whereas roughness increases. However, the effect of a quite large amount of implanted deuterium in the Zr-containing sample [4] also cannot be neglected. Moreover, it is seen from Table 1 that certain increase of n and decrease of k occurred after treatment by deuterium plasma for both sides of AMGN1 sample.

To make sure that the roughness parameters of the mirror surfaces after the action of the deuterium plasma in case of AMGN1 are changed, in addition to angular ellipsometric measurements we have carried out the diagnostics of surface of both samples by AFM (Fig. 2). The AFM results indicate that the impact of deuterium plasma on both surfaces of AMGN1 sample leads to some increase of the roughness parameters such as an average roughness, root mean square height, maximum height etc. compared to AMGN2 sample (Table 2). For example, the value of root mean square height, equals to 1.66 nm for 5 μm × 5 μm AFM image of AMGN2 sample and 3.59 nm for 1 μm × 1 μm AFM image of AMGN1 sample respectively.

It was also found that both values φ_p and $tg\Psi_{min}$ of mirrors from both alloys for mutually perpendicular directions in their own planes coincide within limits of the error of the determination of these parameters. It means that the optical anisotropy of a near-surface layer of the samples both in the initial state and after treatment in the deuterium plasma was not observed being inherent for amorphous surface layer structures.

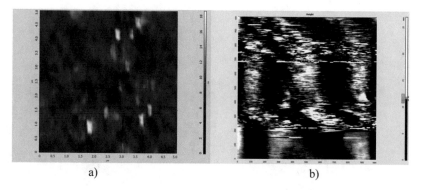

a)   b)

Fig. 2. AFM images of AMGN2 (a) and AMGN1 (b) mirror samples

Table 2. The roughness parameters for face side of AMGN2 and AMGN1 samples

Roughness parameters	AMGN2	AMGN1
Max	18.64 nm	38.28 nm
Min	0 nm	0 nm
Peak-to-peak, Sy	18.64 nm	38.28 nm
Ten point height, Sz	9.33 nm	18.92 nm
Average	7.77 nm	10.97 nm
Average roughness, S_a	1.22 nm	2.68 nm
Second moment	7.94	11.54
Root mean square, S_q	1.66 nm	3.59 nm

The method of the spectral ellipsometry was used for a more detailed study of the optical properties of mirrors after deuterium treatment. In Fig. 3 the dispersion dependences of the optical parameters n, k, ε_1 and σ for one side of AMGN2 and AMGN1 samples in the visible and near infrared range are given. The behavior of $n(E)$, $k(E)$, $\varepsilon_1(E)$ and $\sigma(E)$ curves for another side is similar.

The refraction n and the absorption k indices (Fig. 3(a, b)) are decreased with increasing photon energy E both before and after irradiation by deuterium plasma ions. However the absolute value n of the irradiated AMGN1 sample increases somewhat in the whole investigated spectral range that coincides with angular ellipsometric data.

From Fig. 3(c) it is evident that for both AMGN1 and AMGN2 samples the real part ε_1 of the complex dielectric function has negative values within the spectral range 0.5–3.5 eV. As it is known [5], in metals and metallic alloys the negative contribution to ε_1 comes from the mechanism of intraband acceleration of the charge carriers. In the dependences of the optical conductivity σ (Fig. 3(d)) of both AMGN1 and AMGN2 samples within the investigated spectral range are noticed certain absorption bands consisting of several subbands. One can see that the treatment with deuterium plasma does not significantly change the features in the σ curves but the absolute values of the optical conductivity increase after such processing. Probably the latter is due to the deuterium plasma ions sputtering of a subsurface layer of a certain thickness, which

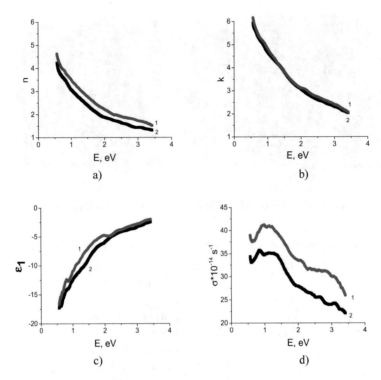

Fig. 3. The dispersion dependences of the optical characteristics n (a), k (b), ε_1 (c) and σ (d) of AMGN1 (1) and AMGN2 (2) samples for the face surface

results in the cleaning of mirrors from an impaired surface layer formed as a result of mechanical polishing.

Since the concentration of zirconium in the investigated BMG mirrors is about half of the total element content, for a comparison of the optical properties of the calculated spectral dependences of σ and ε_1 for Zr are given in Fig. 4(a, b) in according to the data of work [7].

If one compare Fig. 3(d) with Fig. 4(a), it may be concluded that the peculiarities of the σ curves for AMGN1 and AMGN2 samples within the longer wavelength region of the spectrum are due to the presence of large concentration of Zr in these BMGs, but the features of the σ curves at shorter wavelengths (photon energies E of 2.5–3.25 eV) are presumably associated with the presence of other metal components such as Cu. To check whether copper affects the optical spectra, we analyzed similar data obtained for another Zr-based amorphous alloy with grater Cu content.

In Fig. 5(a–d) the dispersion dependences of the optical characteristics n, k, ε_1 and σ of the AM-J_p sample of amorphous metallic alloy $Zr_{48}Cu_{36}Al_8Ag_8$ irradiated by the deuterium plasma are also given in spectral range investigated.

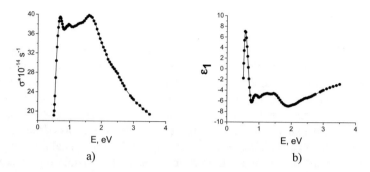

Fig. 4. The spectral dependences of σ (a) and ε_1 (b) for Zr calculated in according to the data of work [7]

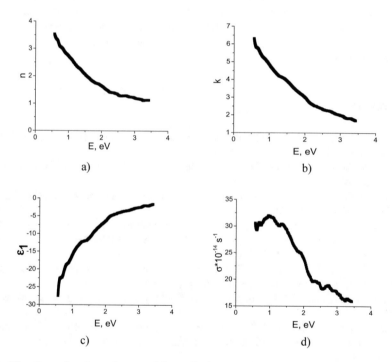

Fig. 5. The dispersion dependences of the optical parameters n (a), k (b), ε_1 (c) and σ (d) of AM-J sample of amorphous alloy-2 exposed in deuterium plasma

One can see that decrease (about of 9 at %) of Zr concentration for $Zr_{48}Cu_{36}Al_8Ag_8$ amorphous alloy and essential increase of Cu concentration by more than twice as compared to $Zr_{57}Cu_{15.4}Al_{10}Ni_{12.6}Nb_5$ alloy slightly change the optical properties in the investigated spectral range.

4 Conclusions

Thus, it was found that the treatment of mirror-like surfaces by the deuterium plasma essentially changes the optical properties of the surface of amorphous samples resulting in measurable differences in the absolute values of ellipsometric characteristic $\text{tg}\,\Psi_{min}$ and the magnitudes of the deviations in the microrelief parameters for the unaffected and the treated surface atomic structure. It was found that ellipsometric parameters obtained in two mutually perpendicular directions in the own plane of the investigated samples relatively their p-direction are not distinguished. The optical anisotropy of the samples both in the initial state and after treatment in the deuterium plasma was not observed, what is inherent for amorphous surface layer structures.

It was revealed that there is slightly difference in behavior of optical properties of $Zr_{57}Cu_{15.4}Al_{10}Ni_{12.6}Nb_5$ and $Zr_{48}Cu_{36}Al_8Ag_8$ amorphous alloys mirrors in the whole spectral range studied (0.5–3.5 eV). For the longer-wavelength range (0.5–2.5 eV) this is probably due to difference in the concentration of Zr, and at shorter wavelengths (photon energies of 2.5–3.25 eV) the difference is presumably associated with elevated concentration of Cu.

References

1. Suryanarayana, C., Inoue, A.: Bulk Metallic Glasses. CRC Press, Taylor & Francis Group, Boca Raton, London, New York (2011)
2. Inoue, A., Wang, X., Zhang, W.: Developments and applications of bulk metallic glasses. Rev. Adv. Mater. Sci. **18**, 1–9 (2008)
3. Hofmann, D.: Bulk metallic glasses and their composites: a brief history of diverging fields. J. Mater. **2013**, 1–8 (2013)
4. Voitsenya, V., Bardamid, A., Balden, M., et al.: On the prospects of using metallic glasses for in-vessel mirrors for plasma diagnostics in ITER. In: Metallic Glasses - Formation and Properties, chap. 7, pp. 135–167. Intechopen, London (2016)
5. Poperenko, L., Stashchuk, V., Shaykevich, I., et al.: Precision Tools and Devices of Optotechnique. Kyiv University, Kyiv (2016). (in Ukraine)
6. Azzam, R., Bashara, N.: Ellipsometry and Polarized Light. Elsevier Science Ltd., Amsterdam (1987)
7. Weaver, J., Krafka, C., Lynch, D., Koch, E.: Optical Properties of Metals I: The Transition Metals, $0.1 \leq hv \leq 500$ eV. Zentralstelle fur Atomkerenergie-Dokumentation, ZAED, Hamburg (1981)

Biotechnology and Environmental Engineering, Electric and Electronic Engineering

Up-Conversion Nanosized Phosphors Based Fluoride for Photodynamic Therapy of Malignant Tumors

Anastasiya M. Dorokhina and Vadim V. Bakhmetyev$^{(\boxtimes)}$

Saint-Petersburg State Institute of Technology,
Moskovsky Prospect, 26, St.-Petersburg, Russia
vadim_bakhmetyev@mail.ru

Abstract. Finely dispersed (particle size less than 100 nm) $NaYF_4:Yb^{3+},Er^{3+}$ and $YF_3:Yb^{3+},Er^{3+}$ phosphors are synthesized using a hydrothermal method. $YF_3:Yb^{3+},Er^{3+}$ phosphor is found to be mostly appropriate for IR radiation based photodynamic therapy (PDT). The efficiency of infrared laser stimulated generation of singlet (active) oxygen by a drug composition containing $YF_3:Yb^{3+},Er^{3+}$ phosphor and "Radachlorin" photosensitizer is characterized.

Keywords: Up-conversion · Nanophosphors · Photodynamic therapy

1 Introduction

Photodynamic therapy (PDT) is a modern promising method of treatment of oncological diseases. Its essence lies in the introduction into the body of a photosensitizer - a substance capable of selectively accumulated in the tumor tissue and producing active oxygen, which destroys the tumor cells. Currently, PDT is widely used in the treatment of skin, lung, larynx, esophagus, stomach and other tumors with superficial localization [1–3].

However, with all the advantages of PDT, its use for the treatment of cavity neoplasms is ineffective. The main problem is the difficulty of bringing light into the organism, since visible light is actively absorbed in the tissues of the body. The necessity of modernizing the PDT method for effective treatment of oncological neoplasms with cavitary localization is obvious.

To expand the use of PDT, it is proposed to create a pharmacological preparation containing, along with a photosensitizer, a colloidal solution of a nanophosphor that converts X-ray or infrared (IR) radiation that easily penetrates through the body tissues into light with a wavelength necessary for the photosensitizer to work [4].

Since photosensitizers are known substances, the most important task is the creation of a nanophosphor. Such nanoparticles should meet the following requirements:

1. Excitation of luminescence should be carried out by radiation from medical therapeutic devices that easily penetrate body tissues (X-rays with a wavelength less than 0.3 Å or IR radiation with wavelengths in the range of 760–1000 nm).

© Springer Nature Switzerland AG 2019
G. Laukaitis (Ed.): INTER-ACADEMIA 2018, LNNS 53, pp. 123–130, 2019.
https://doi.org/10.1007/978-3-319-99834-3_16

2. The emission wavelength must correspond to one of the bands in the photosensitizer absorption spectrum or be as close as possible to ensure the effective work of the preparation. Typically, photosensitizers produced by industry have absorption bands in the "blue" ($\lambda \approx 400$ nm) and "red" ($\lambda \approx 660$ nm) regions of the spectrum.
3. Nanophosphor should be biocompatible, non-toxic and harmless to the body, and also have hydrolytic stability in the blood.
4. The particle size should not exceed 100 nm in order to allow the preparation of a stable colloidal solution suitable for injection into the body [4].

In our previous papers [5–8], phosphors such as $Zn_3(PO_4)_2{:}Mn^{2+}$, $NaBaPO_4{:}Eu^{2+}$, $Y_2O_3{:}Eu^{3+}$, $Y_3Al_5O_{12}{:}Eu^{3+}$ were proposed to improve PDT. The main problem was that all the listed phosphors require, during the synthesis, mandatory high-temperature annealing (more than 800 °C), which leads to the splicing of particles into large agglomerates. To solve this problem, it was suggested to use a hydrothermal method for the synthesis of phosphors, which involves heat treatment of a liquid medium in an autoclave at moderate temperatures below 210 °C. As luminescent materials that can be synthesized by the hydrothermal method, it has been proposed to use fluorides of lanthanides, the advantage of which is the combination of high density, effective X-ray and IR luminescence and non-toxicity. The aim of this work was the hydrothermal synthesis of nanophosphors excited by X-ray and infrared radiation and the study of the properties of synthesized phosphors, including the efficiency of their use in preparations for generation of active oxygen.

2 Experimental Details

2.1 Synthesis of NaYF$_4$:Yb^{3+},Er^{3+} (Aqueous Medium)

In a procedure [9] for the preparation of $NaYF_4{:}25$ mol% Yb^{3+}, 3 mol% Er^{3+} nanocrystals, 1 mmol $RE(NO_3)_3$ (RE = Y, Yb, and Er) was added into 20 ml aqueous solution containing 4 mmol citric acid under vigorous stirring for 1 h. Then 30 ml deionized water containing 6 mmol of NaF (the molar ratio of RE^{3+}/F^- is 1:6) and 30 mmol of $NaNO_3$ was added into the above solution. After being stirred for another 30 min, the as-obtained precursor solution was transferred to a 100 ml Teflon bottle held in a stainless steel autoclave, sealed and maintained at 180 °C for 12 h. After the autoclave was cooled to room temperature, the precipitates were collected by centrifugation (8000 rpm, 15 min) and washed with deionized water for several times.

2.2 Synthesis of YF$_3$:Yb^{3+},Er^{3+} (Organic Medium)

Water-soluble and polyethylene glycol (PEG)-coated $YF_3{:}Yb^{3+},Er^{3+}$ nanoparticles were synthesized by method [10]. 0.2 M solution of $RECl_3$ (RE = Y, Yb, and Er) was added to 44 mL ethylene glycol (EG). Then 2 mmol NaCl was added to above solution and stirred for 30 min. PEG (2 g) (Mw = 2000) was added and sonicated the solution for 15 min. After that, 16 mL of EG containing NH_4F was added to above mixture. The mixture solutions were stirred and sonicated for another 30 min and then transferred into a 100 mL stainless Teflon-lined autoclave and kept at 200 °C for 24 h. The reaction mixture was washed and centrifuged several times with ethanol and bidistilled water to remove other residual solvents and then suspended in bidistilled water for further use.

2.3 Characterization

Investigation of the luminescence spectra of synthesized nanophosphors was carried out with the AvaSpec-3648 spectrofluorometer. To excite the luminescence an infrared laser with an output power of 0.8 W with a wavelength of 980 nm was used. The luminescence intensity was calculated from the area under the curve of the luminescence spectrum. X-ray phase analysis of nanophosphors was carried out on a RigakuSmartLab X-ray diffractometer. The electron micrographs were taken with the TescanVega 3 SBH scanning electron microscope. The production of active oxygen was measured by chemical traps [11] from the change in the optical density of a solution containing a nanophosphor, the Radachlorin photosensitizer, and a substance that is discolored by active oxygen (1,3 diphenylisobenzofuran). The optical density of the solution was measured using a SF-56 single-beam spectrophotometer.

3 Results and Discussion

Figure 1 shows X-ray diffractograms of samples $YF_3:Yb^{3+},Er^{3+}$ and $NaYF_4:Yb^{3+},Er^{3+}$ synthesized in an ethylene glycol medium and in an aqueous medium, respectively. The samples synthesized in the ethylene glycol medium have a nanocrystalline structure (the peaks are relatively wide).

Determination of particle sizes of synthesized phosphors was carried out by scanning electron microscopy. As an example, Fig. 2 shows raster electronic microphotographs of the phosphor $YF_3:Yb^{3+},Er^{3+}$. It can be seen that the size of the phosphor particles does not exceed 100 nm, which makes it possible to prepare a stable colloidal solution suitable for use in photodynamic therapy.

In addition to the particle size, the most important requirement that a phosphor must satisfy for use in PDT is the closeness of its luminescence bands to the photosensitizer absorption bands. Figure 3 shows the absorption spectrum of the industrial photosensitizer "Radakhlorin" produced by the Russian LLC "Rada-Pharma", based on a mixture of sodium chlorin e6 salts, chlorine p6 and purpurin 5. The three most intense bands in the absorption spectrum have maxima with wavelengths of 403 nm, 505 nm and 660 nm.

The luminescence spectra of anti-Stokes nanophosphors synthesized in various media are shown in Fig. 4. Luminescence intensity of the sample synthesized in ethylene glycol is much larger. Presumably, it is the impurity phases of the samples synthesized in water that reduce the intensity and give more intense peak in the region of 580 nm. Thus, the most suitable sample is the synthesis of which was carried out in an ethylene glycol medium.

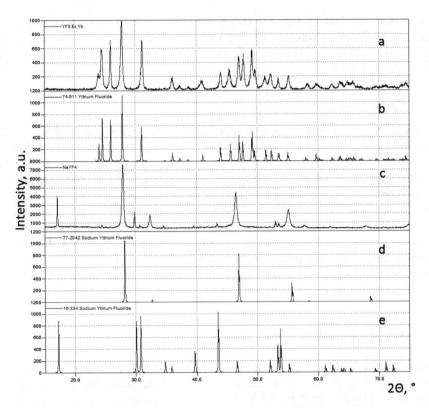

Fig. 1. XRD patterns of the synthesized phosphors: a – YF$_3$:Yb^{3+},Er^{3+} synthesized in ethylene glycol medium; b – YF$_3$ phase (PDF card 74-911); c – NaYF$_4$ synthesized in aqueous medium; d – NaYF$_4$ phase (PDF card 77-2042); e – Na(Y$_{1,5}$Na$_{0,5}$)F phase (PDF card 16-334)

As shown on Fig. 4, the luminescence band with a maximum of 657 nm, related to the electron transition $^4F_{9/2} \rightarrow {}^4I_{15/2}$ [12], has the highest intensity in the spectra of synthesized nanoluminophors, and practically coincides with the absorption band of "Radachlorin" 660 nm. Less intense bands with maxima of 409 nm (the $^2H_{9/2} \rightarrow {}^4I_{15/2}$ transition) and 523 nm (the $^4F_{7/2} \rightarrow {}^4I_{15/2}$ transition) [12] are near the absorption bands of "Radakhlorin" 403 nm and 505 nm, respectively. Thus, in terms of its spectral characteristics, YF$_3$:Yb^{3+},Er^{3+} nanophosphor synthesized in an ethylene glycol medium, is most suitable for use in PDT in combination with "Radachlorin" photosensitizer.

To study the efficiency of generation of active oxygen, a preparation was prepared on the basis of bidistilled water, containing a nanophosphor YF$_3$:Yb^{3+},Er^{3+} at a concentration of 2.0 g/l and a photosensitizer "Radachlorin" at a concentration of 8.75 mg/l. Sodium lauryl sulfate was used as the stabilizer of the colloid solution. The study of the production of active oxygen was carried out by the method of chemical traps [11].

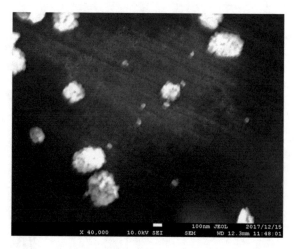

Fig. 2. A scanning electron micrograph of the YF_3:Yb^{3+},Er^{3+} nanophosphor synthesized in an ethylene glycol medium

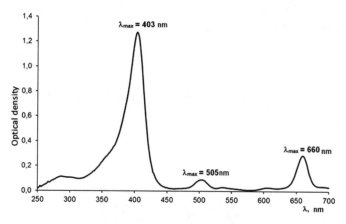

Fig. 3. Absorption spectrum of photosensitizer "Radakhlorin" (production: LLC "Rada-Pharma", Moscow)

To this end, 1,3-diphenylisobenzofuran was added to the colloidal solution containing nanophosphor and "Radachlorin" at a concentration of 0.1 mmol/l. A cuvette containing 3.8 ml of the prepared solution was irradiated with an IR laser with an output power of 0.8 W with a wavelength of 980 nm for 6 h. After irradiation, the phosphor was separated from the solution by centrifugation. Measurement of the optical density was carried out at a wavelength of 326 nm, since this absorption band of 1,3-diphenylisobenzofuran does not coincide with the absorption bands of "Radachlorin". The optical density at a given wavelength as a result of irradiation with an IR laser decreased from 0.64 to 0.60, which corresponds to an oxidation of

0.033 μmol of 1,3 diphenylisobenzofuran and a corresponding amount of active oxygen produced (Fig. 5).

Thus, the efficiency of active oxygen generation using the $YF_3:Yb^{3+},Er^{3+}$ synthesized nanophosphor in combination with "Radachlorin" photosensitizer is 1.9 μmol/MJ of the IR laser energy.

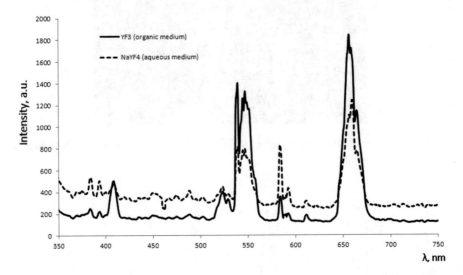

Fig. 4. Luminescence spectra of anti-Stokes nanophosphors synthesized in aqueous medium and in ethylene glycol medium

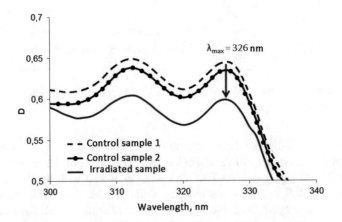

Fig. 5. Absorption spectra, irradiated and control samples

4 Conclusion

As a result of the work done using the hydrothermal method, nanophosphors $NaYF_4:Yb^{3+},Er^{3+}$ and $YF_3:Yb^{3+},Er^{3+}$ were synthesized in two ways (in aqueous medium and ethylene glycol medium). It was found that hydrothermal synthesis in ethylene glycol allows the synthesis of phosphors with particle size less than 100 nm with good crystallinity and suitable for use in PDT. In terms of its spectral characteristics, the $YF_3:Yb^{3+},Er^{3+}$ nanophosphor is the most suitable for use in PDT in combination with the Radachlorin photosensitizer. The efficiency of production of active oxygen by the system "nanophosphor $YF_3:Yb^{3+},Er^{3+}$ - photosensitizer Radakhlorin" under the influence of IR laser with a wavelength of 980 nm was determined.

References

1. Gelfond, M.L.: Photodynamic therapy in oncology. Pract. Oncol. **8**(4), 204–210 (2007)
2. Evtushenko, V.A., Vusik, M.V., Chizhikov, E.A.: Photodynamic therapy of skin relapse cancer with fotoditazin photosensitizer. Russ. Biother. J. **5**(1), 25 (2006)
3. Gelfond, M.L., Barchuk, A.S., Vasilyev, D.V., Stukov, A.N.: Possibilities of photodynamic therapy in oncological practice. Russ. Biother. J. **2**(4), 67–71 (2003)
4. Bakhmetyev, V.V., Sychov, M.M., Orlova, A.I., Potanina, E.A., Sovestnov, A.E.: Nanophosphors for photodynamic therapy of oncological diseases. Nanoindustry **8**, 46–50 (2013)
5. Minakova, T.S., Sychov, M.M., Bakhmetyev, V.V., Eremina, N.S., Bogdanov, S.P., Zyatikov, I.A., Minakova, L.Yu.: The influence of $Zn_3(PO_4)_2$: Mn – luminophores synthesis conditions on their surface and luminescent features. Adv. Mater. Res. **872**, 106–111 (2014)
6. Malygin, V.V., Lebedev, L.A., Bakhmetyev, V.V., Keskinova, M.V., Sychov, M.M., Mjakin, S.V., Nakanishi, Y.: Synthesis and study of luminescent materials on the basis of mixed phosphates. Adv. Intell. Syst. Comput. **519**, 47–54 (2016)
7. Bakhmetyev, V.V., Lebedev, L.A., Vlasenko, A.B., Bogdanov, S.P., Sovestnov, A.E., Minakova, T.S., Minakova, L.Yu., Sychov, M.M.: Luminescent materials on the basis of yttrium oxide and yttrium aluminum garnet used for photodynamic therapy. Key Eng. Mater. **670**, 232–238 (2016)
8. Bakhmetyev, V.V., Minakova, T.S., Mjakin, S.V., Lebedev, L.A., Vlasenko, A.B., Nikandrova, A.A., Ekimova, I.A., Eremina, N.S., Sychov, M.M., Ringuede, A.: Synthesis and surface characterization of nanosized Y_2O_3: Eu and YAG: Eu luminescent phosphors which are useful for photodynamic therapy of cancer. Eur. J. Nanomed. **8**(4), 173–184 (2016)
9. Wang, C., Cheng, X.: Hydrothermal synthesis and upconversion properties of α-$NaYF^4$: Yb^{3+}, Er^{3+} nanocrystals using citric acid as chelating ligand and NaNO3 as mineralizer. J. Nanosci. Nanotechnol. **15**, 9656–9664 (2015)
10. Zeng, S., Tsang, M.-K., Chan, C.-F., Wong, K.-L., Hao, J.: Biomaterials **33**(36), 9232–9238 (2012)

11. Ohyashiki, T., Nunomura, M., Katoh, T.: Detection of superoxide anion radical in phospholipid liposomal membrane by fluorescence quenching method using 1,3-diphenylisobenzofuran. Biochem. Biophys. Acta **1421**, 131–139 (1999)
12. Wang, D.F., Zhang, X.D., Liu, Y.J., Wu, C.Y., Zhang, C.S., Wei, C.C., Zhao, Y.: Hydrothermal synthesis of hexagonal-phase NaYF4:Er, Yb with different shapes for application as photovoltaic up-converters. Chin. Phys. B. **22**, 027801-1–027801-7 (2013)

Multiple-Valued Computing
by Photon-Coupled, Photoswitchable Proteins

Balázs Rakos[1,2(✉)]

[1] Department of Automation and Applied Informatics,
Budapest University of Technology and Economics, Budapest, Hungary
`balazs.rakos@gmail.com`
[2] MTA-BME Control Engineering Research Group, Budapest, Hungary

Abstract. In this work, we discuss the applicability of photon-coupled, photoswitchable proteins in multi-valued computing circuits. The proposed operational principle is based on photoswitchable proteins, capable of switching between more than two forms when subjected to light with well-defined frequencies. The molecules must be able to emit light with specific frequencies, determined by their forms (e.g. fluorescent photoswitchable proteins) in order to enable photo-coupling between neighboring proteins. According to our considerations such protein arrangements are potentially suitable for the realization of low power-consuming, terahertz-frequency, nanoscale, multiple-valued logic circuits.

Keywords: Multi-valued logic · Molecular electronics · Logic circuits
Photoswitchable protein · Photon coupling

1 Introduction

1.1 Multi-valued Logic

While the Boolean system uses two logic values, multi-valued logic (MVL), also known as multiple-valued or many-valued logic, operates with three (ternary [1, 2]) or more truth values. MVL can provide more information content than Boolean logic, and it has superior noise immunity to analog systems [3]. MVL systems require less elementary building blocks (e.g. transistors) and interconnects. Some of its disadvantages are reduced signal-to-noise ratio and speed.

MVL permits problem solving in a more efficient way than the Boolean system, and it has already found various applications such as multi-valued memories, arithmetic circuits, field programmable gate arrays (FPGA) [4, 5]. The StrataFlash flash memory from Intel is an excellent example for its commercial applicability.

MVL systems can be realized with the aid of current-mode CMOS VLSI technology [3, 6, 7], but resonant tunneling diodes (RTD) can also serve as building blocks of such circuits [8]. Molecules may also be considered as potential building blocks of MVL systems [9–11].

© Springer Nature Switzerland AG 2019
G. Laukaitis (Ed.): INTER-ACADEMIA 2018, LNNS 53, pp. 131–136, 2019.
https://doi.org/10.1007/978-3-319-99834-3_17

1.2   Photoswitchable Proteins

Reversibly photoswitchable proteins can be switched between multiple forms (conformations) when they are subjected to light with appropriate frequencies [12]. A subgroup of these molecules, fluorescent photoswitchable proteins can emit light with well-defined frequencies specific to their forms due to fluorescence, when excited with radiation with proper frequency. Their primary application is related to tracking the movement and interaction of proteins.

Several kinds of fluorescent photoswitchable proteins exist [13]. As an example, the Dronpa molecule [14], an engineered variant of the green fluorescent protein, can be switched from its non-fluorescent form to its fluorescent form by radiation with $\lambda_1 = 405$ nm wavelength, and it can be switched back to its fluorescent form by radiation with $\lambda_2 = 488$ nm wavelength.

1.3   Photoswitchable Protein-Based Logic Circuits

Due to their numerous advantageous properties (small size with dimensions on the order of few nm, low power consumption and dissipation, availability with various properties, artificial proteins with desired properties can be designed, cheap, environment friendly, with outstanding self-assembling properties) [15–17], proteins may be excellent building blocks of nanometer-size computing and digital processing architectures of the future.

Although Coulomb-coupled proteins may be suitable for the realization of logic architectures [15–17], photon coupling is an even more promising method for the integration of photoswitchable proteins into computing circuits since it permits more stable, robust operation [18]. In [18], we proposed universal computing architectures based on photon pulse-driven, photon-coupled, hypothetical photoswitchable proteins, potentially suitable for low power consuming and dissipating, terahertz-frequency applications. In [19], we introduced OR and NOR logic gates consisting of photon-coupled, existing, already available photoswitchable proteins.

2   Photon-Coupled, Photoswitchable Protein-Based, Multiple-Valued Computing Architectures

2.1   Operational Principle

We showed in [18, 19] that binary logic circuits are potentially realizable with the aid of photon pulse-driven, photon-coupled, photoswitchable proteins. In the following, we will discuss how such proteins could be used in MVL architectures.

Let us assume that we have got a hypothetical photoswitchable protein (protein engineering may permit the realization of such molecules in the future) with more than two different forms ($form_1$, $form_2$, ...), furthermore it can be switched between the different forms with the aid of radiation with appropriate frequencies (e.g. radiation with f_{12} frequency switches the molecule from $form_1$ to $form_2$, and f_{21} switches it backwards, f_{13} switches the protein from $form_1$ to $form_3$, etc.). Let us also assume that all of the different forms can emit radiation with a certain frequency upon irradiation with light with the appropriate frequency (e.g. $form_1$ can be excited by f_{1e}, which results in an emitted radiation with f_{1g}). The schematic of such operation in the case of a three-form system is displayed in Fig. 1.

Schematic of photoswitching:

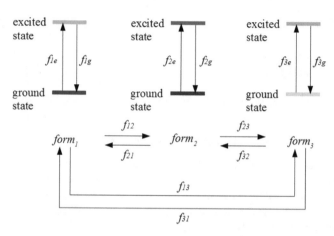

Fig. 1. Schematic of photoswitching of a three-form hypothetical photoswitchable protein

If we assign a certain logic value to each form, we can use such proteins for MVL data storage. The different MVL values can be loaded on the molecule with the application of appropriate light pulses.

In order to create complex MVL circuits from such proteins, they must be coupled together with a suitable method. Photon coupling, just like in the case of binary logic architectures [18], can be utilized in MVL structures, as well.

In the following, we demonstrate the application of this coupling strategy on a simple ternary logic system. Let us consider two, three-form photoswitchable proteins, placed next to each other (see Fig. 2 for their properties). If the molecules are designed in a way that $f^1_{1g} = f^2_{21} = f^2_{31}, f^1_{2g} = f^2_{12} = f^2_{32}, f^1_{3g} = f^2_{23} = f^2_{13}$, then the MVL value of protein$_1$ can be loaded on protein$_2$ with the aid of photon coupling, if protein$_1$ is irradiated by light with $f^1_{1e}, f^1_{2e}, f^1_{3e}$ frequencies. In this way, when protein$_1$ is in $form^1_1$, it emits radiation with f^1_{1g} frequency, which switches protein$_2$ to $form^2_1$, when protein$_1$ is in $form^1_2$, it emits radiation with f^1_{2g} frequency, which switches protein$_2$ to $form^2_2$, and when protein$_1$ is in $form^1_3$, it emits radiation with f^1_{3g} frequency, which switches protein$_2$ to $form^2_3$.

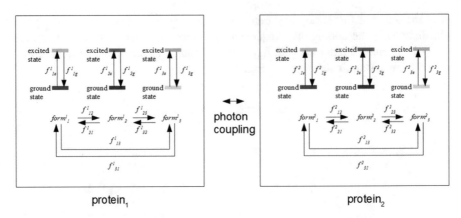

Fig. 2. Schematic of two, photon-coupled three-form photoswitchable proteins

2.2 A Photoswitchable Protein-Based Ternary OR Gate

In order to demonstrate the MVL computing capabilities of proteins described in the previous subsection, we propose a simple, photoswitchable protein-based, ternary OR gate consisting of a single molecule, which can be controlled by well-defined light pulses. The truth table of the two-input ternary OR gate is displayed in Fig. 3, where the three logic values are symbolized with '−', '0', and '+'.

The input logic states can be loaded on the molecule by using light pulses with well-defined frequencies. Before operation the protein must be reset to its $form_1$, which corresponds to the output value of '−'. This can be done with a light pulse consisting of

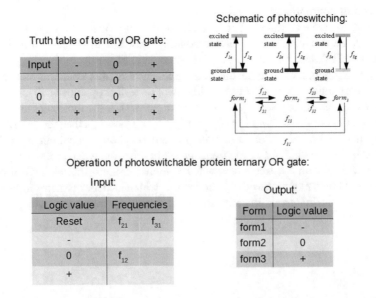

Fig. 3. Truth table and operation of the photoswitchable protein-based ternary OR gate

two frequencies: f_{21}, and f_{31} (if the protein was originally in form$_2$ or form$_3$, this will switch it to form$_1$). After reset the molecule is ready for computation. The input value of '−' corresponds to the absence of light pulses. The '0' value can set by a light pulse consisting of f_{12} frequency, and '+' can be set by frequencies f_{13}, f_{23}.

If the afore-described light pulses are used for setting the two input logic values on the protein, the output value ('−' if the protein is in form1, '0' if it is in form$_2$, and '+' if it is in form$_3$) will correspond to that in the truth table of the ternary OR gate (see Fig. 3).

3 Discussion of the Results

Since reversibly photoswitchable proteins already exist, as it has been demonstrated by several experiments (see introduction), it is reasonable to assume that such molecules with properties required for the desired MVL operations are either naturally available or can be designed with the aid of protein engineering. The dimensions of ordinary proteins are on the order of few nanometers, thereby they can serve as building blocks of nanoscale MVL architectures. The switching speed of presently available photo-switchable proteins is on the order of milliseconds, however, since structural rear-rangements of such molecules can take only few picoseconds [17, 20], with well-designed proteins THz-frequency operations may be possible with such architectures.

In the previous section, we demonstrated that MVL memory storage, MVL signal propagation are theoretically possible with photon pulse-driven, photon-coupled, photoswitchable proteins, which are major requirements for the realization of MVL circuits. We also showed that a simple, two-input ternary OR gate is potentially real-izable with such molecules. These findings indicate that at least certain multiple-valued logic functions, computations may be feasible with the aid of photon-coupled, pho-toswitchable proteins. In order to determine whether universal MVL computing is possible with the afore-described method, further investigations are needed.

4 Conclusion

We showed that it is possible to use photoswitchable proteins, possessing more than two forms for storing multiple-valued logic values. Furthermore, with the aid of photon coupling, it is theoretically viable to propagate MVL signals from one photoswitchable protein to another one. Finally, we proposed a ternary OR gate consisting of a single photoswitchable protein, which can be controlled by appropriate photon pulses. The suggested structures may open the way for low power-consuming and dissipating, terahertz-frequency, nanoscale MVL computing and digital signal processing circuits in the future.

References

1. Hayes, B.: Third base. Am. Sci. **89**, 490–494 (2001)
2. Yoeli, M., Rosenfeld, G.: Logical design of ternary switching circuits. IEEE Trans. Electron. Comput. **EC-14**(1), 19–29 (1965)
3. Current, K.W.: Current-mode CMOS multiple valued logic circuits. IEEE J. Solid-State Circuits **29**(2), 95–107 (1994)
4. Dubrova, E.: Multiple-valued logic synthesis and optimization. In: The Springer International Series in Engineering and Computer Science, vol. 654, pp. 89–114 (2002)
5. Freitas, D.A., Current, K.W.: A quaternary logic encoder-decoder circuit design using CMOS. In: Proceeding International Symposium on Multiple-Valued Logic, pp. 190–195 (1983)
6. Freitas, D.A., Current, K.W.: A CMOS current comparator circuit. Electron. Len. **19**(17), 695–697 (1983)
7. Lin, H.C.: Resonant tunneling diodes for multivalued digital applications. In: Proceeding International Symposium on Multiple-Valued Logic, pp. 188–195 (1994)
8. Jin, N., et al.: Tri-state logic using vertically integrated Si-SiGe resonant interband tunneling diodes with double NDR. Electron Dev. Lett. **25**, 646–648 (2004)
9. de Silva, A.P., James, M.R., McKinney, B.O.F., Pears, D.A., Weir, S.M.: Molecular computational elements encode large populations of small objects. Nat. Mater. **5**, 787–789 (2006)
10. de Silva, A.P., Uchiyama, S.: Molecular logic and computing. Nat. Nanotechnol. **2**, 399–410 (2007)
11. Li, E.Y., Marzari, N.: Conductance switching and many-valued logic in porphyrin assemblies. J. Phys. Chem. Lett. **4**(18), 3039–3044 (2013)
12. Zhou, X.X., Lin, M.Z.: Photoswitchable fluorescent proteins: ten years of colorful chemistry and exciting applications. Curr. Opin. Chem. Biol. **17**(4), 682–690 (2013)
13. Bourgeois, D., Adam, V.: Reversible photoswitching in fluorescent proteins: a mechanistic view. IUBMB Life **64**(6), 482–491 (2012)
14. Ando, R., Mizuno, H., Miyawaki, A.: Regulated fast nucleocytoplasmic shuttling observed by reversible protein highlighting. Science **306**, 1370–1373 (2004)
15. Rakos, B.: Simulation of Coulomb-coupled, protein-based logic. J. Autom. Mob. Robot. Intell. Syst. **3**(4), 46–48 (2009)
16. Rakos, B.: Coulomb-coupled, protein-based computing arrays. Adv. Mater. Res. **2011**(222), 181–184 (2011). https://doi.org/10.4028/www.scientific.net/AMR.222.181
17. Rakos, B.: Modeling of dipole-dipole-coupled, electric field-driven, protein-based computing architectures. Int. J. Circuit Theory Appl. **2015**(43), 60–72 (2015). https://doi.org/10.1002/cta.1924
18. Rakos, B.: Pulse-driven, photon-coupled, protein-based logic circuits. Adv. Intell. Syst. Comput. **519**, 123–127 (2016)
19. Rakos, B.: Photon-coupled, photoswitchable protein-based OR, NOR logic gates. Adv. Intell. Syst. Comput. **660**, 99–103 (2017)
20. Xu, D., Phillips, J.C., Schulten, K.: Protein response to external electric fields: relaxation, hysteresis, and echo. J. Phys. Chem. **100**, 12108–12121 (1996). https://doi.org/10.1021/jp960076a

Investigation of X-Ray Attenuation Properties in 3D Printing Materials Used for Development of Head and Neck Phantom

Jurgita Laurikaitiene[1,2(✉)], Judita Puiso[1], and Evelina Jaselske[1,3]

[1] Faculty of Mathematics and Natural Sciences,
Kaunas University of Technology, Studentu Str. 50, 51368 Kaunas, Lithuania
jurgita.laurikaitiene@ktu.lt
[2] Hospital of Oncology, Hospital of Lithuanian University of Health Sciences
Kauno Klinikos, Volungiu Str. 16, 45433 Kaunas, Lithuania
[3] Hospital of Lithuanian, University of Health Sciences Kauno Klinikos,
Eiveniu Str. 2, 50009 Kaunas, Lithuania

Abstract. 3D printing technologies became an integral part of the medical environment due to their ability to produce relevant copies of human organs and tissues. This feature can be used by developing of radiotherapy/radiology phantoms for patient dose verification thus providing an excellent possibility for individualization of irradiation procedure.

X-ray attenuation properties of four different 3D printing materials (PLA, ASA, PETG and HIPS) thought for phantom construction. Irradiation of samples printed in ZORTRAX300 3D printer was performed in X-ray therapy unit GULMAY D3225; peak voltage of 120 kV was applied. Multipurpose semiconductor detector BARACUDA was used for the assessment of X-ray attenuation in irradiated samples. Experimental results were verified with the results obtained using XCOM data based simulations. It was found, that X-ray attenuation properties of investigated materials were similar to those estimated for thyroid gland, brain, muscle and skin, however differed significantly from attenuation properties in bone and teeth, which are present in the head and neck region and play an important role in attenuation of X-rays in this anatomic region during irradiation procedure.

Keywords: 3D printing materials · Attenuation coefficients · Phantom

1 Introduction

Irradiation of cancer patients using modern radiotherapy techniques requires an accurate, precise and safe dose delivery to the tumour volume, ensuring protection of healthy tissues or/and critical organs. Head and neck patients are always an intrigue issue, as critical organs and healthy tissues are close or in the irradiated area. Therefore improvements in treatment methodology require relevant improvements of methods used for the dose and dose distribution measurements and verification. Some improvement might be realized performing phantom based dose measurements prior to patient's treatment. Various types of phantoms are used in clinical practice [1] or are

© Springer Nature Switzerland AG 2019
G. Laukaitis (Ed.): INTER-ACADEMIA 2018, LNNS 53, pp. 137–143, 2019.
https://doi.org/10.1007/978-3-319-99834-3_18

under development [2–4] with a strong focus on high accuracy of the dose or dose distribution measurements. The creation of inhomogeneous and non-standard geometry phantoms, imitating the real patients anatomical issues (size and shape, critical organs and normal tissues, gender and etc.) [5, 6] are of great interest. Recently 3D printing technologies started to be used for creation of specific phantoms. Using different types of materials it is possible to print out an individualized phantom simulating heterogeneous anatomical structures of the real standard patient, recreating it from the computed tomography (CT) images [4, 7]. The phantoms may differ by their size and the material composition which they are made from and its modification [1, 8–10], but this material must be nearly tissue equivalent and the difference between the density of phantom materials and the human's tissues density has not to exceed 5% [11]. On the other hand even if the mass densities of 3D printed materials are similar enough to tissue densities, the information regarding phantom geometry, correct infill, composition, tissue equivalency, radiation attenuation, absorption and hardness has to be detailed investigated [12–15].

The aim of this work was to investigate X-ray attenuation properties of corresponding tissue equivalent 3D printing materials for fabrication of 3D anthropomorphic phantom prototype.

2 Materials and Methods

2.1 3D Printing Materials

Effective number and mass density of material are parameters which are used for rough estimation of its tissue equivalency. Due to this 3D printing materials having similar density as organs of interest in the Head and Neck region were selected for the investigation (Table 1).

For the assessment of X-ray attenuation properties of PLA, ASA, PETG and HIPS printing materials small $(5 \times 4 \times 2)$ mm^3 experimental plates were printed out using Zortrax300 printer. In order to compare attenuation properties of the same material having different mass density two HIPS PO samples depending on 3D printing pattern and fill density were prepared: one with 50% volume filling and another one - with 90%. Data on experimental sample density is provided in the Table 1. There was some difference observed between the density of original 3D printing filament and 3D printed ready-to-use samples.

Some authors indicated possibility to use PLA [18] or PETG [4] for simulation of bone structures however the density of bone was quite different as compared to the densities of 3D printing materials.

Table 1. Characteristics of 3D printing materials

Materials	Chemical formula	ρ, g/cm^3 [13, 14, 16]	ρ, g/cm^3 (3D printed sample)	Anatomic structure [17]	ρ, g/cm^3
Polylastic acid (PLA)	$C_3H_4O_2$	1.06–1.43	1.012	Thyroid gland	1.045
				Skin	1.09
				Bone? [18]	1.92
Poly(acrylonitrile-co-styrene-co-acrylate) (ASA)	$C_{15}H_{17}NO_2$	1.06–1.1	0.998	Brain	1.05
Glycol-modified polyethylene terephthalate (PETG)	$C_{16}H_{28}O_8$	1.27	1.108	Bone? [4]	1.92
High impact polysterene (HIPS)	$C_{12}H_{16}$	1.04–1.05	0.886 (50%) 0.928 (90%)	Muscle	1.04
Poly(methyl-methacrylate) (PMMA)	$C_5H_8O_2$	1.06	1.05	Neck, trachea	1.05

2.2 Assessment of X-Ray Attenuation

For the assessment of X-ray attenuating properties experimental samples were irradiated in X-ray therapy unit GULMAY D3225 with a conical acrylic applicator (irradiation field – Ø10 cm) using X-ray tube peak voltage of 120 kVp. BARRACUDA multimeter with multi-purpose detector R100B (RTI Electronics) was used for the dose and kVp measurements. Two parameters were calculated: X-ray absorption coefficient (Eq. 1) and mass attenuation coefficient (Eq. 2) of investigated 3D printed materials:

$$B = \left(1 - \frac{D(x)}{D(0)}\right) * 100\%, \tag{1}$$

$$\mu/\rho = \ln[D(0)/D(x)]x * \rho, \tag{2}$$

where D(0) is the entrance dose of the sample and D(x) is the dose beneath the sample, μ is linear attenuation coefficient, x is the thickness of the sample and ρ is the density.

XCOM database [19] was used for the verification of the mass attenuation coefficients obtained from experimental measurements.

3 Results

Evaluation of X-ray attenuating properties of 3D printed materials was performed analysing results of performed dose measurements. At first X-ray absorption properties of 2 mm thick PLA, ASA, PETG and HIPS were evaluated. It was found that PLA possessed the highest X-ray absorption coefficient among all investigated samples however its density was not the highest. This led to suggestion that scattered photons are contributing significantly to the dose D(x) registered in the presence of sample. To support this suggestion XCOM database was used for the simulation of X-ray photon interactions in 3D printing materials. The results of XCOM simulations for each separate material are provided in Fig. 1.

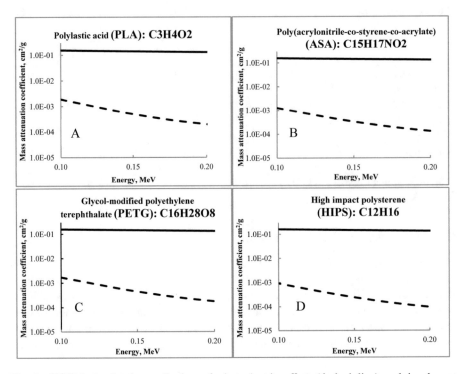

Fig. 1. XCOM simulated contribution of photoelectric effect (dashed line) and incoherent scattering effect (solid line) to total attenuation of X-rays in 3D printed materials

It is evident, that the main contribution to the total X-ray attenuation in 3D printed materials is made by incoherent scattering. It was estimated that this contribution was >95% in the low energy range up to 200 keV. Moreover the reverse of total attenuation coefficient values was observed in this energy range (Fig. 2). It was found that at the energy of ~60 keV materials having the lowest total attenuation coefficient values (HIPS and ASA) are changing to materials with higher attenuation coefficient as compared to other investigated materials. Observed variations of mass attenuation coefficient in X-ray irradiated 3D printed materials might be complementary to explanation why experimentally calculated attenuation coefficient values were showing opposite tendency as compared to those, evaluated using XCOM simulations. Performed calculations revealed that there was no direct dependency between mass density and X-ray attenuation properties in 3D printed materials, and the calculated X-ray total attenuation coefficient values were somewhat higher. These discrepancies between theoretical and experimental calculations indicated clearly the necessity for the evaluation of properties of 3D printed samples prior to start them to use for the

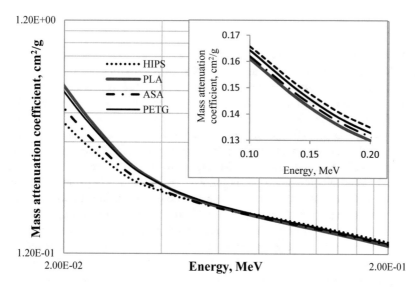

Fig. 2. Variations of XCOM simulated total mass-attenuation coefficient values in 3D printed materials within energy interval of interest

development of medical phantoms. Due to the possible 3D printing pattern and filling density variations, properties of ready-to-use product might differ significantly from the simulated values obtained using XCOM or any other database.

The results of performed investigation are summarized in Table 2.

Table 2. Attenuation characteristics of 3D printed materials

Materials	ρ, g/cm^3	D(0), mGy	D(x), mGy	X-ray absorption coefficient, %	Mass attenuation coefficient, cm^2/g (XCOM)	Mass attenuation coefficient, cm^2/g (experimental)
PLA	1.012	2.154	1.989	7.67	0.1550	0.3937
ASA	0.998	2.156	2.059	4.50	0.1525	0.2306
PETG	1.108	2.150	2.054	4.46	0.1570	0.2061
HIPS 90%	0.928	2.153	2.067	3.99	0.1587	0.2196
HIPS 50%	0.886	2.154	2.078	3.53	-	0.2027

4   Conclusions

Investigation of X-ray attenuating properties in 4 different 3D printed materials (PLA, ASA, PETG and HIPS) has been performed. No direct dependency was found between X-ray attenuating properties of corresponding material and its mass density which is the main parameter used for the estimation of materials equivalency to the biological tissues. This highlighted the fact that in parallel with the mass density at least elemental composition of 3D printed material should be considered since it plays an important role analysing radiation induced scattering effects. Observed discrepancies between theoretical (XCOM based) and experimental data indicated clearly that the X-ray attenuation characteristics of every 3D printed material which is planned for the fabrication of medical phantom should be a priory assessed experimentally considering also 3D printing pattern and infill density in it. Investigation also revealed that X-ray attenuation properties of investigated materials were similar to those estimated for thyroid gland, brain, muscle and skin. However they differed significantly from attenuation properties in bone and teeth, which are present in the head and neck region and play an important role in attenuation of X-rays in this anatomic region during irradiation procedure.

Acknowledgements. This work was partly supported by the research grant No. S-MIP-17-104 of Lithuanian Research Council.

References

1. Winslow, J.F., Hyer, D.E., Fisher, R.F., Tien, Ch.J., Hintenlang, D.E.: Construction of anthropomorphic phantoms for use in dosimetry studies. J. Appl. Clin. Med. Phys. **10**(3), 195–204 (2009)
2. Xu, G.X.: An exponential growth of computational phantom research in radiation protection imaging and radiotherapy: a review of the fifty-year history. Phys. Med. Biol. **59**(18), 233–302 (2014)
3. Carton, A.K., Bakic, P., Ullberg, C., Derand, H., Maidment, A.D.: Development of a physical 3D anthropomorphic breast phantom. Med. Phys. **38**(2), 891–896 (2011)
4. Jeong, H., Han, Y., Kum, O., Kim, Ch., Park, J.: Development and evaluation of a phantom for multi-purpose in intensity modulated radiation therapy. Nucl. Eng. Technol. **4**(43), 399–404 (2011)
5. Yoshitomi, H., Kowatari, M.: Influence of different types of phantoms on the calibration of dosemeters for eye lens dosimetry. Radiat. Prot. Dosimetry **170**(1–4), 199–203 (2016)
6. Giacometti, V., Guatelli, S., Bazalova-Carter, M., Rosenfeld, A.B., Schulte, R.W.: Development of a high resolution voxelised head phantom for medical physics applications. Physica Med. **33**, 182–188 (2017)
7. Adliene, D., Jaselske, E., Urbonavičius, B.G., Laurikaitiene, J., Rudžianskas, V., Didvalis, T.: Development of 3D printed phantom for dose verification in radiotherapy for the patient with metal artefacts inside. IFMBE Proc. **68**(3), 643–647 (2018)
8. Šniurevičiūtė, M., Laurikaitiene, J., Adliene, D., Augulis, L., Rutkūniene, Ž., Jotautis, A.: Stress and strain in DLC films induced by electron bombardment. Vacuum **83**(1), 159–161 (2009)

9. Adlienė, D., Laurikaitienė, J., Tamulevičius, S.: Modification of amorphous DLC films induced by MeV photon irradiation. Nucl. Instrum. Methods Phys. Res. Sect. B **266**(12–13), 2788–2792 (2008)
10. Marcinauskas, L., Grigonis, A., Kulikauskas, V., Valincius, V.: Synthesis of carbon coatings employing a plasma torch from an argon–acetylene gas mixture at reduced pressure. Vacuum **81**, 1220–1223 (2007)
11. International Commission on Radiation Units and Measurements. ICRU 62. Prescribing, recording and reporting photon beam therapy (supplement to ICRU 50). ICRU, Bethesda, MD (1999)
12. Alssabbagh, M., Tajuddin, A., Abdulmanap, M., Zainon, R.: Evaluation of nine 3D printing materials as tissue equivalent materials in terms of mass attenuation coefficient and mass density. Int. J. Adv. Appl. Sci. **4**(9), 168–173 (2017)
13. Savi, M., Potiens, M., Silveira, L.C., Cechinel, C.M., Soares, F.A.P.: Density comparison of 3D printing materials and the human body. IJC Radiol. 1–3 (2017)
14. Kairn, T., Crowe, S.B., Markwell, T.: Use of 3D printed materials as tissue-equivalent phantoms. In: World Congress on Medical Physics and Biomedical Engineering, pp. 728–731. Springer, Toronto (2015)
15. Jeong, S., Yoon, M., Chung, W., Kim, D.: Preliminary study for dosimetric characteristics of 3D printed materials with megavoltage photons. J. Korean Phys. Soc. **67**(1), 189–194 (2015)
16. Alssabbagh, M., Tajuddin, A., Abdulmanap, M., Zainon, R.: Evaluation of 3D printing materials for fabrication of a novel multifunctional 3D thyroid phantom for medical dosimetry and image quality. Radiat. Phys. Chem. **135**, 106–112 (2017)
17. International Commission on Radiation Units and Measurements ICRU 44. Tissue substitutes in radiation dosimetry and measurement, Report 44, Bethesda (1989)
18. Kamomaea, T., Shimizu, H., Nakaya, T., Okudaira, K., Aoyama, T., Oguchi, H., Komori, M., Kawamura, M., Ohtakara, K., Monzen, H., Itoh, Y., Naganawa, S.: Three-dimensional printer-generated patient-specific phantom for artificial in vivo dosimetry in radiotherapy quality assurance. Med. Phys. **44**, 205–211 (2017)
19. NIST National Institute of Standards and Tecnology, Physical Meas. Laboratory Homepage, XCOM: http://www.nist.gov/pml/data/xcom/. Accessed 29 June 2018

Investigation of X-Ray Attenuation Properties in Water Solutions of Sodium Tungstate Dihydrate and Silicotungstic Acid

Laurynas Gilys[1(✉)], Diana Adliene[1] ⓘ, and Egidijus Griskonis[2]

[1] Department of Physics, Faculty of Mathematics and Natural Sciences,
Kaunas University of Technology, Studentu g. 50, 51368 Kaunas, Lithuania
Laurynas.gilys@ktu.edu
[2] Faculty of Chemical Technology, Kaunas University of Technology,
Radvilėnu Pl. 19, 50254 Kaunas, Lithuania

Abstract. Due to its outstanding photon attenuation features lead (Pb) is the most popular material which is used for radiation shielding and for radiation protection of individuals against ionizing radiation. However, Pb is very toxic and can cause serious health problems. Also recycling of lead containing materials is relative complicated. In order to overcome Pb related problems, researchers are looking for lead free materials possessing similar photon attenuation properties as lead and that can be used for the development and fabrication of radiation shielding elements and radiation protection equipment.

In this paper we discuss the investigation results of two tungsten containing water solutions: sodium tungstate dihydrate ($Na_2WO_4 \cdot 2H_2O$) and silicotungstic acid ($H_4SiW_{12}O_{40} \cdot xH_2O$). Aqueous solutions containing different concentrations of tungsten products were fabricated and their X-ray attenuation properties were investigated. Since these solutions were thought for application as the fillers in aquarium type radiation protection screens their lead equivalent was calculated. Radiation protection screen of this type were developed for application in interventional radiology departments. It was found, that the lead equivalent of investigated solutions was dependent on concentration of tungsten compounds. Solutions containing $\geq 30\%$ of sodium tungstate and $\geq 45\%$ of silicotungstic acid indicated lead equivalent ≥ 0.25 mmPb, thus meeting the requirements set for radiation protection equipment.

Keywords: Lead free materials · X-ray shielding
Radiation protection equipment

1 Introduction

Radiation shielding materials are used to protect medical personnel and patients from unwanted radiation during diagnostic procedures [1]. High atomic number materials having also higher density are usually used to attenuate photons or completely absorb X-rays [2–8]. The most popular X-ray shielding material is lead (Pb), because of its high density, high level of stability, flexibility in applications, ease fabrication and availability. Other materials (tungsten, tantalum, gold) have higher density than lead,

© Springer Nature Switzerland AG 2019
G. Laukaitis (Ed.): INTER-ACADEMIA 2018, LNNS 53, pp. 144–149, 2019.
https://doi.org/10.1007/978-3-319-99834-3_19

but Pb can be easily fabricated and is the cheapest of the higher density materials. However, lead has some serious disadvantages: Pb is heavy and very toxic [9, 10]. Due to these reasons the scientific interest for lead free alternative radiation protection materials is increasing. Various polymeric composites containing tungsten [11, 12], bismuth [13], gadolinium [14], tin, barium [15] or their compounds are alternative non-toxic substances that can be used to change toxic lead [16]. Most of the developed composites contain complicated compounds fabrication of which is time consuming and special fabrication technique [17].

The aim of this work was the development, evaluation and comparison of X-ray absorption properties of two different lead free aqueous solutions containing tungsten (sodium tungstate dihydrate and silicotungstic acid) in order to find out their feasibility for application as filler for aquarium type shielding screen used in interventional radiotherapy department [18].

2   Materials and Methods

2.1   Experimental Samples of Tungsten Containing Water Solutions

Two batches of experimental samples were prepared following sample preparation procedure described elsewhere [12]: water solutions containing 5 different concentrations (5%, 15%, 25%, 35%, 42%) of sodium tungstate dihydrate ($Na_2WO_4 \cdot 2H_2O$) and water solutions containing 5 different concentrations (15%, 30%, 45%, 60%, 79%) of silicotungstic acid ($H_4SiW_{12}O_{40} \cdot 24H_2O$). Selection of concentrations was limited by solubility of tungsten containing compound in water. Prepared transparent solutions were poured into standard cuvettes ($10 \times 10 \times 42$) mm^3, tightly closed and left to set for 24 h. Photographs of as prepared samples are provided in Fig. 1.

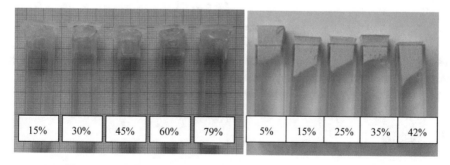

Fig. 1. Photograph of lead free aqueous solutions containing different concentrations of silicotungstic acid (left) and sodium tungstate (right)

2.2   Experimental Set Up for X-Ray Attenuation Measurements

Measurements of X-ray attenuating properties of experimental samples containing tungsten were performed following internationally accepted guidelines [19–21] and using experimental set up shown in Fig. 2 [11].

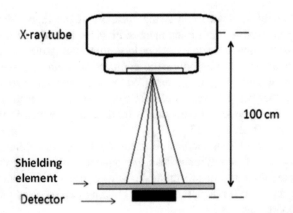

Fig. 2. Experimental set up for the evaluation of X-ray attenuation properties of shielding elements (cuvettes with tungsten containing water solutions)

Cuvettes filled with tungsten containing water solution were placed in direct irradiation field of diagnostic X-ray machine MULTIX PRO. Tube voltage was set to 120 kV. Distance between X-ray tube focal spot and cuvettes was kept 100 cm during the investigations. "Barracuda" multimeter with multi-purpose detector R100B (RTI Electronics) was used for the dose and kVp measurements (Fig. 3).

In general, two parameters were evaluated: X-ray transmission coefficient, B (x) (Eq. 1) and lead equivalent, x (Eq. 2).

$$B(x) = \frac{D(x)}{K(0)} \tag{1}$$

$$x = \frac{1}{\alpha\gamma} \ln\left(\frac{B^{-\gamma} + \frac{\beta}{\alpha}}{1 + \frac{\beta}{\alpha}}\right) \tag{2}$$

where $K(0)$ is the air kerma measured at the certain position without sample; $D(x)$ is the dose, measured beneath the sample which was placed on the top of the detector; α, β and γ are fitting parameters, depending on the voltage applied (kVp).

3 Results

The lead equivalent values evaluated for experimental samples are provided in the Table 1.

Table 1. Lead equivalent values for tungsten containing experimental samples

$H_4SiW_{12}O_{40}$ concentration in water solution, %	15	30	45	60	79
Lead equivalent, mmPb	0.164	0.185	0.251	0.362	0.576
Na_2WO_4 concentration in water solution, %	5	15	25	35	42
Lead equivalent, mmPb	0.066	0.143	0.225	0.290	0.339

Total X-ray mass attenuation coefficients of investigated materials were obtained using XCOM data based [22] simulations. Variations of lead mass attenuation coefficient with energy applied were included for the comparison between attenuation properties of Pb and experimental tungsten containing samples.

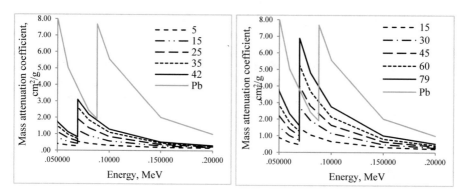

Fig. 3. Variations of X-ray attenuation with energy: mass attenuation coefficient for solutions containing different concentrations of sodium tungstate dihydrate (left) and of silicotungstic acid (right). Concentrations are indicated as numbers in figure legends

It is known that K-edge of tungsten is present at 69.52 keV, while K-edge of lead - at 88.01 keV. This might be of advantage, since due to atomic structure of the W, tungsten containing materials can more effective absorb X-rays in a specific energy range between both 70 and 80 keV. Performed simulations revealed that experimental water solutions containing high concentrations of silicotungstic acid (45%, 60% and 79%) recorded better established X-ray attenuation properties in this specific energy range, as compared to X-ray attenuation in pure lead. However, sodium tungstate containing water solutions indicated only modest increase of the mass attenuation in the energy range mentioned above. X-ray attenuation was linearly dependent on tungsten compound concentrations in water solutions (Fig. 4).

It is known that X-ray attenuating properties of materials are directly linked to lead

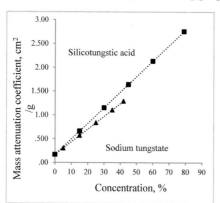

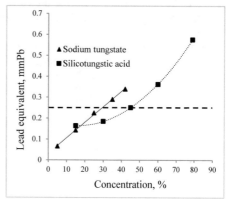

Fig. 4. Concentration dependent variations of total mass attenuation coefficient at 100 keV

Fig. 5. Concentration dependent variations of lead equivalent

equivalent values. It should be noted that there is a requirement [20] of having ≥ 0.25 mmPb lead equivalent for radiation protection shields used in interventional radiology departments. Analysis of lead equivalency for differently concentrated tungsten containing solutions has shown, that in general, lead equivalent was higher for sodium tungstate solution and required value of 0.25 mmPb for 10 mm thick cuvette was achieved at lower concentrations ($\geq 30\%$), as compared to $\geq 45\%$ concentrations of silicotungstic acid solution. However, the increase of sodium tungstate concentration in water solution was limited due to the saturation limit of $\leq 42\%$ of this material in water. The required lead equivalent value of 0.25 mmPb was achieved at 45% concentration of silicotungstic acid dissolved in water. Lead equivalent was increasing linearly with the increasing concentration of silicotungstic acid in water solution, however with reduced cost effectiveness and was limited only by atomic structure of material itself (Fig. 5).

4 Conclusions

Investigation of two tungsten containing water solutions has shown that sodium tungstate and silicotungstic acid aqueous solutions can be used in the construction of radiation protection equipment (aquarium type shielding screens) thus replacing lead containing constructions in the interventional radiology energy range. The minimum requirement for lead equivalent of 0.25 mmPb was achieved with the 30% sodium tungstate and 45% silicotungstic acid concentration solutions. However due to the water solubility limit of sodium tungstate at 42%, the highest achieved lead equivalent value was 0.34 mmPb. The highest achieved lead equivalent value of 0.58 mmPb for silicotungstic acid solutions was limited only by atomic structure of material itself. In the energy range 70–80 keV silicotungstic acid containing solutions were absorbing X-rays better than pure lead.

References

1. McCaffrey, J.P., Shen, H., Downton, B., Mainegra-Hing, E.: Radiation attenuation by lead and nonlead materials used in radiation shielding garments. Med. Phys. **34**(2), 530–537 (2007)
2. Aral, N., Nergis, F.B., Candan, C.: The X-ray attenuation and the flexural properties of lead-free coated fabrics. J. Ind. Text. **47**(2), 252–268 (2017)
3. Singh, A.K., Singh, R.K., Sharma, B., Tyagi, A.K.: Characterization and biocompatibility studies of lead free X-ray shielding polymer composite for healthcare application. Radiat. Phys. Chem. **138**, 9–15 (2017)
4. Aral, N., Banu Nergis, F., Candan, C.: An alternative X-ray shielding material based on coated textiles. Text. Res. J. **86**(8), 803–811 (2016)
5. Nambiar, S., Osei, E.K., Yeow, J.T.: Polymer nanocomposite-based shielding against diagnostic X-rays. J. Appl. Polym. Sci. **127**(6), 4939–4946 (2013)
6. Gaikwad, D.K., Obaid, S.S., Sayyed, M.I., Bhosale, R.R., Awasarmol, V.V., Kumar, A., Pawar, P.P.: Comparative study of gamma ray shielding competence of WO_3-TeO_2-PbO glass system to different glasses and concretes. Mater. Chem. Phys. **213**, 508–517 (2018)

7. Shik, N.A., Gholamzadeh, L.: X-ray shielding performance of the EPVC composites with micro-or nanoparticles of WO_3, PbO or Bi_2O_3. Appl. Radiat. Isot. **139**, 61–65 (2018)
8. Chang, L., Zhang, Y., Liu, Y., Fang, J., Luan, W., Yang, X., Zhang, W.: Preparation and characterization of tungsten/epoxy composites for γ-rays radiation shielding. Nucl. Instrum. Methods Phys. Res. Sect. B **356**, 88–93 (2015)
9. Honigsberg, H., Speroni, K.G., Fishback, A., Stafford, A.: Health care workers' use and cleaning of X-ray aprons and thyroid shields. AORN J. **106**(6), 534–546 (2017)
10. Pulford, S., Fergusson, M.: A textile platform for non-lead radiation shielding apparel. J. Text. Inst. **107**(12), 1610–1616 (2016)
11. Adlienė, D., Griškonis, E., Vaičiūnaitė, N., Plaipaite-Nalivaiko, R.: Evaluation of new transparent tungsten containing nanocomposites for radiation protection screens. Radiat. Prot. Dosimetry **165**(1–4), 406–409 (2015)
12. Gilys, L., Griškonis, E.: Investigation of optical and X-ray absorption properties of polymeric composites containing sodium tungstate. In: Proceedings of the 12th International Conference on Medical Physics, 5–7 November, Kaunas, Lithuania, pp. 108–110 (2015)
13. Maghrabi, H.A., Vijayan, A., Deb, P., Wang, L.: Bismuth oxide-coated fabrics for X-ray shielding. Text. Res. J. **86**(6), 649–658 (2016)
14. Shamshad, L., Rooh, G., Limkitjaroenporn, P., Srisittipokakun, N., Chaiphaksa, W., Kim, H. J., Kaewkhao, J.: A comparative study of gadolinium based oxide and oxyfluoride glasses as low energy radiation shielding materials. Prog. Nucl. Energy **97**, 53–59 (2017)
15. Chanthima, N., Kaewkhao, J., Limkitjaroenporn, P., Tuscharoen, S., Kothan, S., Tungjai, M., Limsuwan, P.: Development of BaO–ZnO–B_2O_3 glasses as a radiation shielding material. Radiat. Phys. Chem. **137**, 72–77 (2017)
16. Verma, S., Sanghi, S.K., Amritphale, S.S.: Development of advanced, non-toxic, X-ray radiation shielding glass possessing barium, boron substituted kornerupine crystallites in the glassy matrix. J. Inorg. Organomet. Polym Mater. **28**(1), 35–49 (2018)

Single Molecule Force Spectroscopy on Collagen Molecules Deposited on Hydroxylated Silicon Substrate

Alexandra Besleaga and Lucel Sirghi[✉]

Iasi Plasma Advanced Research Center (IPARC), Faculty of Physics,
Alexandru Ioan Cuza University of Iasi,
Blvd. Carol I Nr. 11, 700506 Iasi, Romania
lsirghi@uaic.ro

Abstract. Development of single-molecule force spectroscopy instruments as atomic force microscope (AFM), optical tweezers and magnetic tweezers, allows for manipulation of individual molecules and measurements of intermolecular forces with pico Newton resolution. In all these techniques the molecules are physically or chemically bound to larger bodies, which can be moved with high spatial precision. In typical force spectroscopy experiments performed by AFM, the AFM tip is pushed for some time to a substrate in order to pick up a molecule from the substrate surface. Binding of the molecule to the tip and substrate is probed by the occurrence of the characteristic entropic force-extension pattern of the molecule on the force-displacement curve observed during the tip retraction form the substrate.

In the present work collagen type I molecules isolated from rat tail were deposited on silicon wafers. To enhance formation of hydrogen bonds between collagen molecules and either silicon AFM probes and substrates, the AFM probes and substrates were cleaned and hydroxylated in negative glow plasma of a dc discharge in low-pressure air. Then, the collagen molecules were deposited by imbedding the hydroxylated silicon wafers in aqueous solutions of collagen molecules for 2 h. AFM images of dried collagen samples showed a complete coverage of substrates with a thick layer of collagen molecules. A sample consisting in isolated collagen molecules deposited on freshly cleaved mica is used for comparison. Single molecule stretching experiments were made on large numbers of collagen molecules picked up from either hydroxylated silicon or mica substrates. The entropic force-extension curves obtained in experiments were fitted using the worm like chain model of chain molecules to determine the contour length, persistence length and binding force values. Dispersion of the values of these parameters obtained in a statistically significant number of measurements are discussed on base of statistical variation in molecule binding sites and contribution of molecule-substrate interaction forces to the measured single-molecule stretching force.

Keywords: Single-molecule force spectroscopy · Collagen type I
Atomic force microscopy · Plasma hydroxylation

© Springer Nature Switzerland AG 2019
G. Laukaitis (Ed.): INTER-ACADEMIA 2018, LNNS 53, pp. 150–159, 2019.
https://doi.org/10.1007/978-3-319-99834-3_20

1 Introduction

Recent development of instruments capable of simultaneous measurements of tiny forces and displacements enables the investigation of the mechanical properties of single molecules in controlled physico-chemical conditions [1, 2]. Such investigations have been performed to study various biomolecules as DNA [3], RNA [4], proteins [5] and polysaccharide [6], as well as of other long chain polymer molecules [7]. The most used techniques of manipulating single molecules include atomic force microscopy (AFM) [8], optical tweezers [9] and magnetic tweezers [10]. In all these techniques single-molecules are physically or chemically bound to larger bodies, which can be moved with high spatial precision. For the case of AFM technique, one end of a molecule is bound to the tip of an AFM probe and the other end is bound to a flat substrate. To do this, the molecules to be studied are bonded on a substrate where from they are picked up by the AFM tip. The tip and substrate surfaces are chemically functionalized in order to bind the ends of the studied molecules through specific chemical interactions. Nonspecific physical adsorption of molecules on tip and substrate surfaces may be also used to bind molecules, but the binding sites of molecules are not controlled. In some experiments the molecules form a monolayer grafted on the substrate [11]. There are also experiments when the molecules are picked up by the AFM tip from their natural substrate, as in cases of collagen from tendons [12] and polysaccharides from a bacterial surface [13]. In all these experiments, the AFM tip is pushed to the sample surface for some time in order to bind a molecule to its surface. Binding of a molecule to the tip is probed by the occurrence of the characteristic entropic force-extension pattern of the molecule on the force-displacement curve observed during the tip retraction form the substrate. However, in these experiments the exact positions of the molecule binding sites on the AFM tip and sample are usually unknown. Nevertheless, the procedures that are routinely used for fitting single-molecule force-extension curves obtained in AFM measurements consider that the molecule binding position on the substrate is exactly underneath the tip apex. Moreover, the intermolecular forces (Van der Waals, double layer, etc.) between the long chain molecule and substrate are neglected. Given the experimental conditions described above, it is reasonable to assume that there is a certain lateral (perpendicular to the AFM tip displacement direction) distance, r, between the molecule binding sites on the tip and substrate, respectively. Figure 1(a) illustrates the case when the AFM tip is in contact with sample surface and $r \neq 0$. The value of r for each particular molecule stretching experiment is usually unknown. The present work gives an analysis of the consequences of this unknown variable for the case of single-molecule stretching experiments performed by AFM. We simulated AFM force-distance curves affected by occurrence of finite values of r and then fitted these force-distance curves with the force-extension formulae predicted by worm like chain molecule models to extract the characteristic parameters (contour and persistence lengths) of the molecules. Neglecting of the lateral distance results in important underestimations of the fitting parameters and molecule-surface binding force. This result was confirmed by single-molecule stretching experiments performed by AFM in de ionized water on collagen type I molecules randomly picked up by a hydroxylated AFM tip from silicon or mica substrates. As a rule, the persistence length and contour length of collagen molecules determined in these experiments were much smaller than the values reported in literature for single-molecule stretching experiments and molecular dynamic simulations on collagen type I molecules [14]. For a set of a large number of single-molecule extension experiments, when identical chain molecules are randomly picked up by the AFM tip, dispersions in the values of fitting parameters and binding force are observed due to the dispersion of r values and action of the intermolecular forces between collagen molecules and substrates.

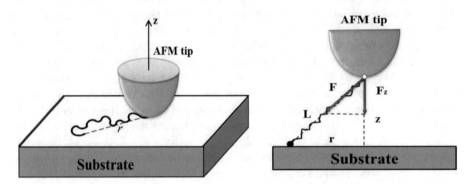

Fig. 1. Left: sketch of the AFM tip, molecule, and sample substrate in the initial position of a single-molecule extension experiment performed by AFM. The AFM tip is in contact with the substrate in order to pick up one end of the molecule, while the other end of the molecule is bound to the substrate. The end-to-end distance of the chain molecule on the substrate is the lateral distance, r, which affects the force-distance curve measured when the tip is retracted from the substrate. Right: sketch that represents the cross section of the AFM tip and sample during AFM tip retraction along the vertical direction (z).

2   Theoretical Considerations

Since long chain molecules are made up of numerous monomers, they are treated by means of statistical physics. Stretching of a chain molecule requires energy (and force) to reduce the number of possible chain configurations and thus decrease the molecule entropy. If the internal energy of the molecule does not change, the force required to extend the molecule along certain direction is purely entropic. In this case the dependence of force, F, on the molecule extension distance, L, can be obtained from mathematical models of chain molecules. The most used models are free joint chain and worm like chain (WLC) [15]. It has been proved that the mechanical properties of collagen molecules are well described by WLC model. This model takes into account the elastic energy stored by the chain curls. Bending of a small portion of the chain with the length ΔL by a small angle $\Delta\theta$ requires the elastic energy

$$\Delta E = \frac{L_p}{\Delta L} \cdot \frac{k_B T}{2} \cdot \Delta\theta^2 \tag{1}$$

where L_p is the persistence length, a parameter that characterize the flexibility of the chain molecule. The more flexible a chain molecule is, the shorter the persistence length is. There is no exact analytical formula for the force-extension dependence of a WLC molecule. An approximate interpolation formula [16],

$$F(L) = \frac{k_B T}{L_p} \cdot \left[\frac{1}{4 \cdot (1 - L/L_c)} + \frac{L}{L_c} - \frac{1}{4} \right] \tag{2}$$

has been proven to work well especially at either low ($L/L_c < 0.2$) or large ($L/L_c > 0.8$) molecule extension values. A better approximate formula has been proposed by Bouchiat [17]:

$$F(L) = \frac{k_B T}{L_p} \cdot \left[\frac{1}{4 \cdot (1 - L/L_c)} + \frac{L}{L_c} - \frac{1}{4} + \sum_{i=2}^{i=7} \left(\frac{L}{L_c} \right)^i \right] \tag{3}$$

with $a_2 = -0.5164228$, $a_3 = -2.737418$, $a_4 = 16.07497$, $a_5 = -38.87607$, $a_6 = 39.49944$, and $a_7 = -14.17718$. The accuracy of this formula is 0.01% and it is used in the present work to fit AFM force-extension curves.

As shown in the left side of Fig. 1, because of the finite lateral distance ($r \neq 0$) between molecule binding sites on the AFM probe and sample, respectively, the molecule stretching direction is changing during the AFM tip retraction. The AFM tip has only one degree of freedom, which is along z axis, and the cantilever vertical deflection determine only the component along this axis, F_z, of the molecule extension force, F. Thus the force measured in the experiment is

$$F_z = F \cdot \frac{z}{\sqrt{z^2 + r^2}} \tag{4}$$

and the tip-substrate distance, z, is related to the molecule extension length, L, by:

$$z = \sqrt{L^2 - r^2} \tag{5}$$

Therefore, the force-distance curves obtained in single-molecule extension experiments performed by AFM are affected by the transformations described by Eqs. (4) and (5). As result, fitting of the force extension curves by Eq. (3) provides correct values for L_p and L_c only for the case $r = 0$, when $F_z = F$ and $z = L$. The measurements of the binding force of the molecule to either AFM tip or substrate are also affected by r. Thus, while the molecule bound breaks up at certain extension force value, F_b, the measured force is:

$$F_b' = F_b \cdot \frac{z_b}{\sqrt{z_b^2 + r^2}} \tag{6}$$

where z_b is the tip-sample break-up separation distance and F_b' is the measured value of z component of the binding force, F_b. Therefore, for $r \neq 0$, the effect of lateral stretching is an important underestimation of binding force.

The effect of r on the force extension curves can be eliminated if the value of r is known, by applying inverse transformation equations:

$$F = F_z \cdot \frac{\sqrt{z^2 + r^2}}{z} \tag{7}$$

and

$$L = \sqrt{z^2 + r^2}. \tag{8}$$

These transformations can be applied to the experimental force-distance curves prior to their fitting with the theoretical chain molecule model. Another approach is to take the distance r as a third fitting parameter and to look for the best fit of the experimental data with Eq. (3) modified by the transformations described by Eqs. (4) and (5). Unfortunately, the nonlinear formulae used by this approach become too complex and the correlation between fitting parameters (especially between r and L_p) becomes too strong to discriminate a good value for r. Finally, it is worth to remark that if the contour length of the stretched molecule is known, the inverse transformations (4) and (5) with trial values of r may be applied to the experimental force-distance curves until the best fitted value of L_c corresponds to the known L_c value, in which case the best fitted value of L_p can be trusted.

3   Experiment

Vial of collagen type I from rat tail (4 mg/ml) in 0.1 Mol solution of acetic acid was procured from Sigma Aldrich and diluted in deionized water. To obtain substrates covered by isolated collagen molecules, freshly cleaved mica substrates (agar scientific) were imbedded in collagen solution with low concentration (5 µg/ml) for 2 h. For collagen deposited on hydroxylated silicon substrates it was used a higher collagen concentration (10 µg/ml) and a longer imbedding time (12 h). Before deposition, the silicon substrates were cleaned and hydroxylated by treating them in the negative glow plasma of a glow discharge in air at low pressure (the pressure, discharge voltage and current intensity were around 0.4 Torr, 400 V and 10 mA, respectively) for 10 min. After depositions, the samples were gently cleansed with deionized water and dried in nitrogen flow.

The dried samples were loaded on the AFM machines (Solver Pro from NT-MDT, Russia) and scanned in tapping mode with a silicon AFM probe (NSG 30 from Mikromasch Inc.) with sharp tip (curvature radius smaller than 10 nm). Topography images of dried samples of collagen on mica and silicon substrates are shown in Fig. 2. Individual molecules are clearly visible on mica substrate. On silicon substrate, it is difficult to distinguish individual molecules because the substrate is uniformly covered by a layer (around 10 nm in thickness) of collagen molecules.

For single molecule stretching experiments, the collagen samples were loaded on the liquid cell of the AFM machine and the cell was filled with dionized water. The experiments were performed with contact silicon AFM probes (CSG 11 from Mikromasch) with the resonance frequency of 8.8 kHz, force constant of 0.01 nN/nm and tip

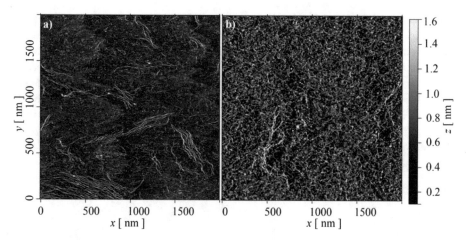

Fig. 2. Topography images (2 μm × 2 μm) of collagen molecules deposited on (a) freshly cleaved mica and on (b) plasma hydroxylated silicon

curvature radius of 10 nm. Force-probe displacement curves were performed on arrays of 20 × 20 positions homogeneously distributed on an area of 2 μm × 2 μm of sample surface. Each force curve were performed with a vertical displacement speed of the AFM probe of 1 μm/s. The force-displacement curves that showed characteristic single-molecule stretching events were then processed by a homemade software to determine L_c, L_p and F_b. Figure 3 shows two force-extension curves recorded in single-molecule stretching experiments in deionized water for collagen on mica substrate. It appears that the force-extension curve is more affected by lateral distance between molecule pining points in the case (a) than in the case (b), when the fitting parameters are more close to the values reported in literature. As analyzed in the theoretical considerations, the lateral distance between the molecule pining points and interaction between molecule and substrate determine smaller values of L_p and L_c. Since in these experiments the binding sites of collagen molecules are not controlled, we processed statistically the results found by analyzing a set of about 100 single-molecule stretching experiments for the cases of collagen molecules picked up by an hydroxylated AFM tip from mica and hydroxylated silicon substrates. The histograms of the values of L_p, L_c and detachment force found in these experiments are presented in Figs. 4 and 5. As expected, due to various positions of molecule binding sites and interaction between molecules and substrate during stretching experiments, values of L_p were much smaller than the values reported in literature for collagen type I molecules. The maximum value of L_p was around 2 nm for molecules picked up from silicon substrate and around 10 nm for the molecules picked up from mica substrate (Fig. 4a). This is explained by low density of collagen molecules on mica substrate, which favors a weaker interaction of molecules that are stretched with other molecules on the substrate. On the other hand, molecules picked up from silicon substrate interact with many neighboring molecules on substrate, fact which results in an apparently smaller value of L_p. The values of L_c are also smaller than the maximum contour length of collagen molecules because either large lateral distance between binding sites or binding sites that are not at the ends of collagen molecules. The most probable binding force between AFM tip and collagen molecules is due to formation of hydrogen bonds between hydroxyl groups on the silicon surface of hydroxylated AFM probes and oxygen or nitrogen atoms in the collagen molecules. Most probably, the available nitrogen or oxygen atoms for formation of hydrogen bonds are at the ends or at defects in the secondary structure of collagen molecules. The histogram in Fig. 5 show a dispersion of the values of binding force measured at the detachment of molecules from the either tip or substrate in the range 40–160 pN with maximum of occurrence at 70 pN. These values are close to the values reported for the hydrogen bonds between hydroxylated AFM tips and phospholipids molecules [18].

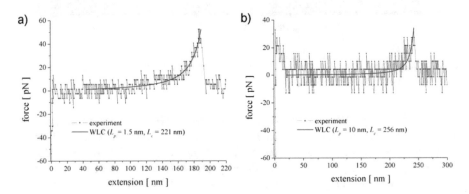

Fig. 3. Example of force-extension curves fitting with WLC model for single-molecule stretching experiments performed in deionized water with hydroxylated AFM tips on collagen type I molecules picked up from mica substrate

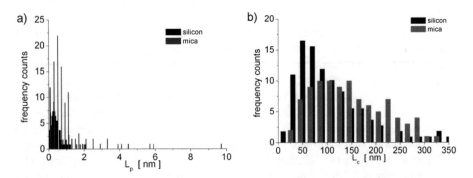

Fig. 4. Histograms of (a) L_p and (b) L_c values as they were determined by fitting of the experimental data recorded in 100 single-molecule stretching experiments performed in deionized water on collagen type I molecules picked up randomly by an hydroxylated AFM tip from silicon substrate and mica, respectively

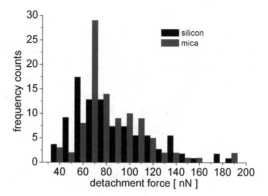

Fig. 5. Histograms of detachment force values determined in 100 single-molecule stretching experiments performed in deionized water on collagen type I molecules picked up randomly by an hydroxylated AFM tip from silicon substrate and mica, respectively

4 Conclusion

If force-extension data obtained for a large number of single-molecule extension experiments performed by AFM are fitted with a chain molecule model, dispersions in the fitting parameters and binding force values are usually observed. In these experiments, the AFM tip picks up identical chain molecules randomly from a natural or artificial substrate. Although the exact positions of the molecule binding sites on the tip and sample in single-molecule extension experiments performed by AFM are unknown, analyses of experimental data usually assume that the molecule binding positions are on the tip apex and underneath the AFM tip apex on the substrate, respectively. Moreover, the interaction between molecules and substrate during stretching is neglected. More realistic assumptions are that there is a certain unknown lateral (perpendicular to the AFM tip moving direction) distance between molecule binding sites on the tip and substrate, respectively, and, while the molecule is stretched, there is interaction between the molecule and substrate.

In this work we show that the dispersion of the experimental data obtained in statistical sets of single-molecule events can be attributed to various positions of molecule binding sites to AFM tip and substrate, respectively, and to the interaction between stretched molecules and substrate (or other neighboring molecules). In the case of WLC molecules as collagen type I, the occurrence of a lateral distance between molecule binding sites on the AFM tip and substrate, respectively, results in smaller values of persistence length, contour length and binding force. We analyzed statistically the values of persistence length, contour length and binding force obtained in sets of about 100 single-molecule stretching events performed on collagen type I molecules picked up randomly from mica or hydroxylated silicon substrates. The sample consisting in collagen molecules on mica was used as reference. To prepare this sample, collagen molecules were deposited with a very low surface density on mica in order to insure that the AFM tip picks up a single molecule and the interaction between stretched molecule and substrate is weak. The sample consisting in hydroxylated silicon was covered by a layer of collagen molecules, in which case a molecule picked up by the AFM tip from this substrate interacts with many other neighboring collagen molecules. Histograms of persistence length and contour length showed a dispersion of data with maximum values closed to the results reported by other authors for single-molecule stretching experiments of collagen type I molecules. Binding force values were also dispersed and showed maximum occurrence probability around 70 pN, a value that can be attributed to formation of hydrogen bonds between hydroxylated surface of silicon AFM tip and collagen molecules.

References

1. Janshoff, A., Neitzert, M., Oberdörfer, Y., Fuchs, H.: Force spectroscopy of molecular systems—single molecule spectroscopy of polymers and biomolecules. Angew. Chem. Int. Ed. **39**(18), 3212–3237 (2000)
2. Lavery, R., Lebrun, A., Allemand, J.F., Bensimon, D., Croquette, V.: Structure and mechanics of single biomolecules: experiment and simulation. J. Phys.: Condens. Matter **14** (14), R383 (2002)
3. Smith, S.B., Cui, Y., Bustamante, C.: Overstretching B-DNA: the elastic response of individual double-stranded and single-stranded DNA molecules. Science **271**(5250), 795–799 (1996)
4. Abels, J.A., Moreno-Herrero, F., Van der Heijden, T., Dekker, C., Dekker, N.H.: Single-molecule measurements of the persistence length of double-stranded RNA. Biophys. J. **88** (4), 2737–2744 (2005)
5. Rief, M., Gautel, M., Oesterhelt, F., Fernandez, J.M., Gaub, H.E.: Reversible unfolding of individual titin immunoglobulin domains by AFM. Science **276**(5315), 1109–1112 (1997)
6. Rief, M., Oesterhelt, F., Heymann, B., Gaub, H.E.: Single molecule force spectroscopy on polysaccharides by atomic force microscopy. Science **275**(5304), 1295–1297 (1997)
7. Bemis, J.E., Akhremitchev, B.B., Walker, G.C.: Single polymer chain elongation by atomic force microscopy. Langmuir **15**(8), 2799–2805 (1999)
8. Fisher, T.E., Marszalek, P.E., Oberhauser, A.F., Carrion-Vazquez, M., Fernandez, J.M.: The micro-mechanics of single molecules studied with atomic force microscopy. J. Physiol. **520** (1), 5–14 (1999)
9. Wang, M.D., Yin, H., Landick, R., Gelles, J., Block, S.M.: Stretching DNA with optical tweezers. Biophys. J. **72**(3), 1335–1346 (1997)
10. Strick, T.R., Allemand, J.F., Bensimon, D., Bensimon, A., Croquette, V.: The elasticity of a single supercoiled DNA molecule. Science **271**(5257), 1835–1837 (1996)
11. Al-Maawali, S., Bemis, J.E., Akhremitchev, B.B., Leecharoen, R., Janesko, B.G., Walker, G.C.: J. Phys. Chem. B **105**, 3965–3971 (2001)
12. Gutsmann, T., Hassenkam, T., Cutroni, J.A., Hansma, P.K.: Sacrificial bonds in polymer brushes from rat tail tendon functioning as nanoscale velcro. Biophys. J. **89**(1), 536–542 (2005)
13. Camesano, T.A., Abu-Lail, N.I.: Biomolecules **3**, 661–667 (2002)
14. Buehler, M.J., Wong, S.Y.: Entropic elasticity controls nanomechanics of single tropocollagen molecules. Biophys. J. **93**, 37–43 (2007)
15. Strobl, G.: The Physics of Polymers. Springer, Berlin (1996)
16. Bustamante, C., Marko, J.F., Siggia, E.D., Smith, S.: Science **265**, 1599–1600 (1994)

17. Bouchiat, C., Wang, M.D., Allemand, J.F., Strick, T., Block, S.M., Croquette, V.: Estimating the persistence length of a worm-like chain molecule from force-extension measurements. Biophys. J. **76**(1), 409–413 (1999)
18. Apetrei, A., Sirghi, L.: Stochastic adhesion of hydroxylated atomic force microscopy tips to supported lipid bilayers. Langmuir **29**, 16098–16104 (2013)

Modification of Polyetheretherketone Surface by Argon, Oxygen and Nitrogen Plasma for Dentistry Application

Zivile Rutkuniene[1](✉), Monika Pervazaite[1], and Gediminas Skirbutis[2]

[1] Kaunas University of Technology, Studentu Str. 50-220, Kaunas, Lithuania
zivile.rutkuniene@ktu.lt
[2] Faculty of Odontology, Lithuanian University of Health Sciences,
Kaunas, Lithuania

Abstract. Due to the chemical and radiation resistance, strength, inertia and biocompatibility properties, synthetic polycrystalline thermoplastic polymer polyetheretherketone (PEEK) is used in implantology and dentistry. Successful application in medicine requires proper surface wetting properties. Implantology intended to provide a hydrophilic, better attachment of cells to the implant surface. Meanwhile, different dental prostheses are required to be hydrophobic. The changes of contact angle, surface energy, and morphology of the polyetheretherketone surface after treatment in nitrogen, oxygen and argon plasma were investigated in this work. The dentine and gingiva PEEK MED 98H14 samples were cleaned for 10 min. with ultrasound in alcohol surrounding before plasma treatment which influenced 18°–23° decreasing of surface contact angle. It was noted that dentine PEEK MED 98H14 was more sensible for the argon and oxygen plasma treatment when ion energy is higher (500 eV) and surface of those samples becomes more hydrophobic. Repeating of experiments at the same conditions with gingiva PEEK MED 98H14 showed the opposite results – contact angle of surface decreased and samples become more hydrophilic. Treatment of both materials with nitrogen plasma resulted the decrease of contact angle by 5°.

Keywords: Polietheretherketone · Plasma treatment · Surface energy

1 Introduction

Polyetheretherketone (PEEK) is a white, organic linear, aromatic polycrystalline thermoplastic derived from polyurethane (PAEC) materials family. This polymer can have crystalline or amorphous phase. There are three main types of PEEK: pure, with carbon or SiO_x glass additions [1]. PEEK is soluble at 335–343 °C temperature and its maximum operating temperature in the air reaches 250 °C. This material has a tensile force of 90–100 MPa, Young's modulus of 3–4 GPa [2], and low friction. In addition, it is hydrophobic, resistant to vapors and to sea water, and have a contact angle of 80°–90° [2] or 70° [3] (depending on the modification of the PEEK surface). PEEK is widely known for its biocompatibility and close to human bones properties. The polymer retains its properties at high temperatures due to its high mechanical (hardness,

© Springer Nature Switzerland AG 2019
G. Laukaitis (Ed.): INTER-ACADEMIA 2018, LNNS 53, pp. 160–164, 2019.
https://doi.org/10.1007/978-3-319-99834-3_21

durability, and etc.) and chemical resistance, even in highly aggressive chemical environment [4]. It is suitable for re-modification in the temperatures above 170° [2].

Considering mechanical and physical properties similar to human bones, PEEK is a potential candidate to be used in implantology and other dentistry applications. Improving the better attachment of cells to the surface of PEEK dental implants without compromising their mechanical properties is a major challenge. Many polymers have a low surface energy which makes the implantology complicated. At the same time, it is important that the materials used for producing removal prostheses have the lowest possible surface energy in order to avoid the rapid deterioration and the formation of dental overgrown [5, 6]. Therefore, it is important to find technologies reducing surface hydrophobicity of PEEK implants. Active gas plasma bombardment is a possible technology changing the surface energy of PEEK polymer [7]. The aim of this research was to investigate the dependence of surface energy on ion energy and nature of biological neutral gas. Three different biological neutral gases (oxygen, nitrogen and argon) were used in those experiments and ion energy was changed from 250 eV to 500 eV.

2 Experimental Setup

Dentine and gingiva PEEK MED 98H14 (Dental Direct Handels GmbH) were used in experiments, both having the same sample size of 98.5 × 14 mm. DD PEEK has a density of ∼ 1.5 g/cm^3, flexural strength force of 1696 MPa, elongation factor ≥ 5%, and softening temperature (Vicat) of 305 °C [8].

Surface modification was proceeded with oxygen, nitrogen, and argon gas plasma in an asymmetric diode type PEVCD 13.6 MHz radio frequency reactor. Pretreatment pressure was 6.1 × 10^{-2} mTorr, working pressure −1 × 10^{-1} mTorr, duration of processes −10 min, ion energy was changed from 250 eV to 500 eV.

Contact angle measurements were repeated five times using 2 μl distilled water drop. Photo pictures were analyzed with Microcapture program. Temperature of surrounding during the measurements was 22 °C. Surface energy was calculated using Owens/Wendt method from the collected data of contact angle measurements:

$$\gamma_L \cdot (1 + cos\theta) = 2 \cdot \left(\sqrt{\gamma_S^D \cdot \gamma_L^D} + \sqrt{\gamma_S^P \cdot \gamma_L^P} \right), \tag{1}$$

where γ_L – surface tension (mN/m), γ_S^D - dispersive component of surface energy (mJ/m^2), γ_L^D – dispersive component of surface tension (mN/m), γ_L^P - polar component of surface tension, γ_S^P - polar component of surface energy (mJ/m^2), θ – contact angle (°).

Additional surface morphology measurements and analysis of post-treatment changes were performed by fixing the image of the surface with an Eclipse LV100ND motorized microscope with epsopropic (tungsten)/diascopic (halogen) illumination.

3 Results and Discussions

In order to investigate the influence of plasma treatment on the surface properties, Dentine PEEK MED 98H14 samples were bombarded with Ar, N_2 and O_2 plasma at different ion energy of 250 eV and 500 eV. The conditions of treatment and main surface characteristics are presented in Table 1. It was obtained that Dentine PEEK MED 98H14 surface before treatment and cleaning has a contact angle of 69°, but it decreased by 18°–20° after ultrasound cleaning in 70% alcohol liquid (Table 1).

Table 1. Dependence of surface properties of Dentine PEEK MED 98H14 on ion energy E_i. Here θ_n is a contact angle after the treatment of Dentine PEEK MED 98H14 and before cleaning, θ is a contact angle of the treated area, γ is a surface energy, γ_S^P and γ_S^D are polar and dispersive components of surface energy, respectively

Plasma	E_i (eV)	γ (mJ/m^2)	γ_S^P (mJ/m^2)	γ_S^D (mJ/m^2)	θ (°)	θ_n (°)
Ar	500	6.07	4.31	0.20	125.6	42.4
O_2	500	12.02		6.14	102.0	44.2
O_2	500	30.6		24.70	83.6	53.8
Ar	250	57.00		51.19	67.0	52.4
O_2	250	58.88		53.00	66.0	48.2
N_2	500	78.78		72.91	55.4	50.2

Measurements of contact angle showed that higher energy of argon and oxygen plasma ions bombardment significant decreased the surface energy and made the surfaces to become more hydrophobic. The higher energy of ion bombardment has the influence of surface roughness and contact angle, as a consequence. Optical microscope pictures proved that more even surface is obtained when contact angle of samples was high (see Fig. 1). In lower energy case, the energy of ion is not enough for full sputtering of surface, therefore, the surface roughness is higher and the contact angle

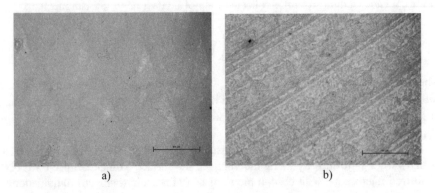

a)   b)

Fig. 1. Optical microscope pictures of dentine PEEK MED 98H14 surfaces after treatment in: (a) O_2 plasma - ion energy 500 eV, contact angle 102°; (b) Ar plasma – ion energy 250 eV, contact angle 67°.

lower. As shown in Table 1, treatment of surface with nitrogen plasma did not change significantly the contact angle and, simultaneously, the surface roughness and chemical composition of surface did not change as well.

Experiments that showed the best results of surface hydrophobicity of dentine PEEK MED 98H14 samples (Ar ir O_2 plasma, ion energy - 500 eV) was repeated with gingiva PEEK MED 98H14. Gingiva PEEK MED 98H14 before cleaning had lower contact angle than dentine (69° and 61°, respectively). Plasma treatment for gingiva PEEK MED 98H14 gave opposite results - contact angle decreased to 24.1° when it was bombarded with Ar plasma, and to 25.2° when it was bombarded with O_2 plasma. The results are summarized in Table 2.

Table 2. Dependence of surface properties of Gingiva PEEK MED 98H14 on ion energy E_i. Here θ_n is a contact angle after the treatment of Gingiva PEEK MED 98H14 and before cleaning, θ is a contact angle of the treated area, γ is a surface energy, γ_S^P and γ_S^D are polar and dispersive components of surface energy, respectively.

Plasma	E_i (eV)	γ (mJ/m^2)	γ_S^P (mJ/m^2)	γ_S^D (mJ/m^2)	θ (°)	θ_n (°)
Ar	500	141.76	4.31	137.45	21.4	41
O_2	500	137.06		132.75	25.2	

4 Conclusions

Differences of contact angle and surface energy for dentine and gingiva PEEK MED 98H14 materials before the treatment are not significant and give similar contact angle values of 69° and 61°, respectively. However, the reaction to the plasma treatment is opposite. It was found that the bombardment of oxygen and argon high-energy plasmas (500 eV) increased contact angle for Dentine PEEK MED and decreased for gingiva PEEK MED. Therefore, the combination of these two materials could be used for manufacturing of removal dentist prostheses and could help to achieve the desired results, i.e. gingiva PEEK MED 98H14 treated with high energy argon or oxygen plasma is more suitable for prosthesis interaction with biological material and the part of prostheses that interact with oral liquids can be manufactured from dentine PEEK MED 98H14. The results show that interacted surfaces can be treated by biological neutral gas plasma for the prevention of the dental overgrow formation.

References

1. Harting, R., Barth, M., Bührke, T., Pfeferle, R.S., Petersen, S.: Functionalization of polyetheretherketone for application in dentistry and orthopedics. Bionanomaterials **18**(1), 1–12 (2017)
2. Najeeb, Sh., Zafar, M.S., Khurshid, Z., Siddiqui, F.: Applications of polyetheretherketone (PEEK) in oral implantology and prosthodontics. J. Prosthodont. Res. **60**(1), 12–19 (2016)

3. Dufils, J., Faverjon, F., Heau, Ch., Donnet, Ch., Benayoun, S., Valette, S.: Evaluation of variety of a-C: H coatings on PEEK for biomedical implants. Surf. Coat. Technol. **313**, 96–106 (2017)
4. Steven, M., Kurtz, J.: Biomaterials in trauma, orthopedic, and spinal implants. Biomaterials **28**, 4845–4869 (2017)
5. Almasi, D., Iqbal, N., Sadegi, M., Sudin, I., Rafiq, M., Kadir, A., Kamarul, T.: Preparation methods for improving PEEK's bioaktivity for orthopedic and dental application: a review. Int. J. Biomater. **1**, 1–12 (2016)
6. Jha, S., Bhowmik, S., Bhatnagar, N., Bhattacharya, N.K., Deka, U., Iqbal, H.M., Benedictus, S.R.: Experimental investigation into the effect of adhesion properties of PEEK modified by atmospheric pressure plasma and low pressure plasma. Appl. Polym. **118**(1), 173–179 (2010)
7. Chua, P.K., Chen, J.Y., Wang, L.P., Huang, N.: Plasma-surface modification of biomaterials. Rep. Rev. J. **36**, 143–206 (2002)
8. PX DENTAL SA homepage. https://www.pxdental.com/de/product/dd-peek-med-98-polymere#&gid=null&pid=1. Accessed 23 April 2018

Thermal Convection of a Phase-Changing Fluid

Takashi Mashiko[1]($\boxtimes$), Yoji Inoue[1], Yuki Sakurai[1],
and Ichiro Kumagai[2]

[1] Shizuoka University, 3-5-1 Johoku, Naka-ku, Hamamatsu 432-8561, Japan
mashiko.takashi@shizuoka.ac.jp
[2] Meisei University, 2-1-1, Hodokubo, Hino, Tokyo 191-8506, Japan

Abstract. To investigate the effect of phase transitions of the fluid on thermal convection, we are conducting experiments of thermal convection of the mixture of a thermosensitive gel and water. The gel absorbs water and swells below a certain critical temperature, while it discharges water and contracts above the critical temperature. The swelling ratio, as well as the critical temperature, can be controlled, which enables us to investigate the effect of phase transition in a series of experiments, where the critical temperature is set between the top and bottom boundary temperatures and the swelling ratio is systematically changed. In flow visualization, we have observed temporally stable and spatially fixed low-velocity regions which are surrounded by distinct high-velocity regions. Such a phenomenon is not observed in thermal convection of a single-phase fluid like water or air, and peculiar to the phase-changing fluid. Also, we have measured the rheological properties of the fluid and found, for example, the shear-thinning behavior, with which we try to explain the observed convection behaviors.

Keywords: Thermal convection · Phase transition · Thermosensitive gel

1 Introduction

Thermal convection, which occurs in a fluid cooled from the top and heated from the bottom, is a ubiquitous phenomenon and has been studied for a long time. In particular, vast knowledge has been accumulated about the Rayleigh-Bénard (R-B) convection occurring in a fluid between horizontal solid plates; state transitions, heat transfer, flow structures, fluctuation properties, and so on. In real situations, however, thermal convection sometimes involves factors that are not considered in the traditional R-B convection, and some have been dealt with recently. For example, thermal convection in spherical shells, modeling the Earth configuration, was numerically studied [1], and the periodic motion of a plate introduced on a free surface of thermal convection, reminiscent of the continental drift, was experimentally studied [2].

The phase transition of the fluid is also a factor that sometimes appears in real phenomena like the Earth's mantle convection but has not been considered in the traditional R-B convection. It is expected that phase transitions of the fluid, which are accompanied by drastic changes of the fluid properties and latent heat transfer, interact with and affect the convection behaviors. Experimental studies have been done to

© Springer Nature Switzerland AG 2019
G. Laukaitis (Ed.): INTER-ACADEMIA 2018, LNNS 53, pp. 165–168, 2019.
https://doi.org/10.1007/978-3-319-99834-3_22

investigate thermal convection with phase transitions of the fluid, but are limited in number and to particular transitions like the one between ice and water [3]. To investigate the effect of phase transitions of the fluid on thermal convection in detail, we have started a novel type of experiment, in which the transition characteristics can be systematically changed.

2 Experiment

We use the mixture of the gel of poly(N-Isopropyl-Acrylamide) (PNIPAM) and pure water as the test fluid of thermal convection. The PNIPAM gel swells by absorbing water below a critical temperature T_c, while it contracts by discharging water above T_c. The phase transition is reversible and its time scale is much shorter than that of convection. The swelling ratio (the mass of the swelled gel divided by that of contracted gel) R, as well as T_c, can be controlled by varying the conditions of PNIPAM synthesis. Setting T_c between the top and bottom temperatures leads to phase transitions of the fluid under thermal convection. Also, changing R enables a series of experiments, from the case corresponding to a single-phase fluid ($R \sim 1$) to the case involving two distinct phases ($R \gg 1$). The synthesized PNIPAM is grinded into grains of tens of micrometers, and mixed with pure water at 0.5 wt%.

The apparatus to be filled with the fluid is a cell of 20 cm × 20 cm × 1 cm (Fig. 1). The top copper plate is cooled by circulating water and the bottom copper plate is heated by an electrical heater. The temperatures of both plates, T_{top} and T_{bot}, are measured by thermocouples ($T_{top} < T_c < T_{bot}$). The convection behavior is observed through the transparent acrylic sidewalls.

A small amount of tracer particles are seeded in the fluid, which is illuminated by a laser of 532 nm to visualize the flow. We use fluorescent particles (FLUOSTAR, EBM Co., Ltd.) of 15 μm, which yield fluorescence of wavelength 580 nm when illuminated by the laser. By using a low-pass filter of cut-off wavelength 550 nm, we can selectively obtain particle images without optical disturbance caused by the light source. From the images recorded by a camera in front of the apparatus, we obtain two-dimensional velocity field by the particle image velocimetry (PIV).

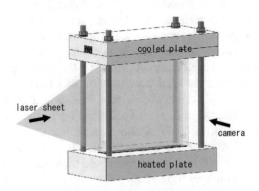

Fig. 1. Experimental apparatus

3 Results and Discussion

We show in Fig. 2 a typical flow image (top left) and the corresponding velocity distribution calculated by the PIV method (top right), obtained in an experiment using a gel of $T_c = 50$ °C and $R = 70$, and at $T_{top} = 10$ °C and $T_{bot} = 60$ °C. We notice some low-velocity regions such as an elliptical area in the top and a triangle area in the left, which are surrounded by high-velocity regions. These regions are spatially fixed and stable for at least 1000 s, as shown in the time series of velocity magnitude (Fig. 2, bottom) at points A and B of the top right image.

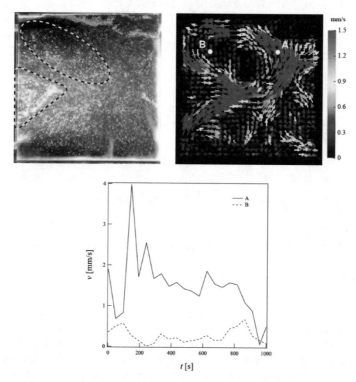

Fig. 2. Visualized flow (top left), corresponding velocity distribution (top right), and time series of velocity magnitude at points A and B (bottom)

This phenomenon presumably results from the multi-phase property of the fluid. In thermal convection of a single-phase fluid like water or air, such a distinct separation of high-velocity and low-velocity regions is not observed. Also, this observation is intriguing, being reminiscent of the mantle convection of the Earth, where the flow passages are rather fixed and they may be oblique (*i.e.*, hotspots are not necessarily right below volcanoes).

To interpret and understand the observed phenomena, it is necessary to grasp the fluid properties. We have tried some rheological measurements. One of the measured properties is the viscosity μ of the mixture of a gel of $T_c = 40$ °C and $R = 150$ with water. Figure 3(left) shows μ as a function of the temperature T, measured at the shear rate of $D = 5$ s^{-1}. We see that μ decreases with increasing T, with a rapid decrease at around T_c. This suggests that, roughly speaking, the fluid is a viscous gel near the cold top plate and a less viscous suspension (PNIPAM particles in water) near the hot bottom plate in thermal convection.

Figure 3 (right) shows μ as a function of the shear rate D at different temperatures. We see again that the higher T, the lower μ, and that μ exhibits hysteresis. Also, we notice the shear-thinning behavior, i.e., the larger D, the lower μ. This seems consistent with the observed phenomenon in the convection experiment; fluidized parts are likely to remain fluidized, while non-fluidized parts remain non-fluidized.

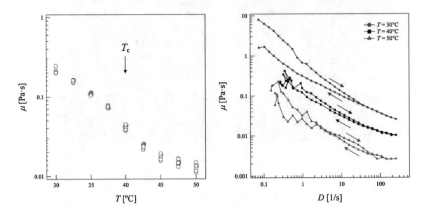

Fig. 3. Fluid viscosity as a function of temperature (left) and shear rate (right)

4 Conclusion

We started an experiment of thermal convection using a thermosensitive gel, and have observed the segregation of the fluid into high-velocity regions and low-velocity regions, which is peculiar to a multi-phase fluid. We also measured rheological properties of the fluid and found a shear-thinning behavior, which is consistent with the observation. Now we are conducting a series of experiment, where the phase-changing property is systematically varied, and more detailed measurement of the rheological properties.

References

1. Gastine, T., Wicht, J., Aurnou, J.M.: Turbulent Rayleigh-Bénard convection in spherical shells. J. Fluid Mech. **778**, 721–764 (2015)
2. Zhang, J., Libchaber, A.: Periodic boundary motion in thermal turbulence. Phys. Rev. Lett. **84**, 4361–4364 (2000)
3. Tankin, R.S., Farhadieh, R.: Effects of thermal convection currents on formation of ice. Int. J. Heat Mass Transf. **14**, 953–956 (1971)

A Nanoantenna-MIM Diode-Lens Device Concept for Infrared Energy Harvesting

Muhammad Fayyaz Kashif[1(✉)] and Balázs Rakos[1,2]

[1] Department of Automation and Applied Informatics,
Budapest University of Technology and Economics, Budapest, Hungary
fayyaz.kashif@aut.bme.hu
[2] MTA-BME Control Engineering Research Group, Budapest, Hungary

Abstract. In this paper, we introduce an antenna-based energy harvesting device for generating electricity from infrared radiation. The proposed device is based on nanoantennas, terahertz rectifying diodes and concentrator lenses. Diodes integrated together with antennas can transform the energy of electromagnetic radiation into electricity by rectifying the high frequency currents induced in the antennas. We consider an array of diode coupled nanoantennas over a suitable substrate. A layer of micro lenses is placed on top of the antenna array to focus the incident light. This can considerably boost the conversion efficiency of the antenna array. A theoretical description of the proposed device is presented. Based on the existing experimental results, conversion efficiency of $\sim 77\%$ is estimated for the 10 μm infrared radiation.

Keywords: Energy harvesting · Nano-antenna · MIM diode · Micro-lens
Rectenna · Infrared radiation · Solar cell

1 Introduction

Sunlight is an exceptional renewable energy source. Solar energy reaches the earth in the form of electromagnetic radiation. Visible and infrared radiation constitute the major portion of the solar spectrum being approximately 39 and 52%, respectively. As a heated object, earth also reemits the infrared radiation back into the atmosphere in the wavelength range of 8–14 μm with a peak at 10 μm [1].

The photovoltaic (PV) cell is the most mature technology regarding direct conversion of sunlight into electricity. It is a quantum mechanical semiconductor device and has a maximum conversion efficiency of about 30%. PV cells have a fundamental limitation related to the bandgap of the semiconductor material and only photons of visible light and near-infrared radiation can be captured and converted to electricity [2]. They are expensive due to cost of materials and fabrication procedures. Another major drawback of these devices is that they can operate only in daytime when visible light is available. This means that even with 100% efficient PV cells the energy from only 50% of the solar spectrum can be harnessed and the potential of solar energy is underutilized by this technology. Therefore, it is necessary to look for more innovative and efficient techniques so that the energy of the entire solar spectrum could be captured.

© Springer Nature Switzerland AG 2019
G. Laukaitis (Ed.): INTER-ACADEMIA 2018, LNNS 53, pp. 169–176, 2019.
https://doi.org/10.1007/978-3-319-99834-3_23

Since sunlight is composed of electromagnetic radiation, it is possible to convert it into electricity by a well-designed antenna. The idea of utilizing antenna for solar energy harvesting was first proposed in 1972 by Baily [3]. The antenna can receive the electromagnetic radiation and can confine them to the feed gap. A diode is integrated together with the antenna to convert high frequency alternating current to direct current. The combination of an antenna with a diode is called a rectenna, which is the abbreviation for rectifying antenna. Metal-Insulator-Metal (MIM) diodes, which consist of two metallic layers sandwiched by one or more thin (few nanometers) insulating layers, are considered suitable candidates for the rectification of high frequency induced current by the visible and infrared radiation [4]. The conceptual block diagram of a rectenna is shown in Fig. 1. With the development of nanotechnology, it is now possible to fabricate nanostructures. Sophisticated nanofabrication techniques like electron beam lithography allow the precise fabrication of lateral structures with sub-micron feature sizes. A number of rectenna devices have been actually fabricated using modern fabrication techniques [5]. Current research is focused on the optimization of the nonlinearity and resistance of the MIM diode, which determines the AC-DC conversion efficiency. Gadalla et al. [6] have demonstrated the fabrication of low resistance, 500 Ω, asymmetrical MIM diode (Gold and Copper) with 67 nm × 67 nm contact area and 0.7 nm oxide thickness by electron beam lithography and atomic layer deposition techniques. Promising results have been achieved to realize symmetric and asymmetric MIM diodes with high curvature at zero bias [7, 8], however, additional research is required to optimize the fabrication and the materials.

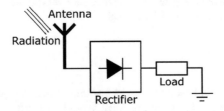

Fig. 1. Block diagram of a rectenna device

In this paper, we introduce a nanoantenna-MIM diode-micro lens device for energy harvesting from ambient infrared radiation. We estimate the conversion efficiency of the proposed device by taking into account the existing measured data of fabricated nanoantenna coupled MIM diodes and micro lenses. Briones et al. [9] examined the conversion efficiency of solar rectennas. However, they did not take into account the effect of concentrator lenses. To the best of our knowledge, this is the first analysis in which the effect of concentrator lenses on the efficiency of solar rectennas is considered. Since the performance of these structures is not limited by heating significantly, the application of concentrators in these devices is expected to be even more beneficial than in the case of semiconductor PV cells. A maximum theoretical efficiency of 86.8% is possible if thermodynamic effects are taken into account [10].

2 Nanoantenna-MIM Diode-Lens Device

We propose an array of nanoantennas integrated with MIM diodes on a suitable substrate as a chip. A layer of micro lenses can be patterned over it. The intensity and the power density of the radiation on the antenna can be increased by focusing the incoming radiation to the focal spot of lens where the devices are placed. In this way, the effective aperture of the antenna and the current level at the output the antenna can be enhanced. A scheme of the so called nanoantenna-MIM diode-micro lens device is displayed in Fig. 2.

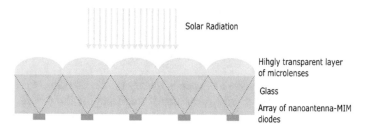

Fig. 2. A conceptual scheme of the proposed nanoantenna-MIM diode-micro lens device

2.1 Nanoantenna-MIM Diode

An MIM diode is a quantum mechanical device. It consists of two metallic electrodes separated by one or more very thin insulating layers. The diode is called symmetric if the metals are made of the same material. For different metals, the diode is called asymmetric. Asymmetric diodes do not require any external biasing and electrons can tunnel through the oxide barrier due to difference between the work functions of the metals. The required oxide thickness is less than 4 nm for tunneling current [11]. MIM diodes can rectify high frequency alternating current to direct current due to non-linear current-voltage characteristics caused by tunneling mechanism. Since tunneling is a very fast process, on the order of femtoseconds, MIM diodes can operate at infrared frequencies.

A simple equivalent circuit model of the nanoantenna-MIM diode is shown in Fig. 3. The antenna is modeled by a high frequency voltage source and a series impedance which depends upon the geometry and material of the antenna. The diode is represented by the capacitance C_d in parallel with a nonlinear resistance R_d [12].

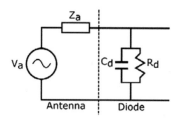

Fig. 3. Equivalent circuit diagram of an antenna-coupled MIM diode

The amplitude of the voltage drop across the diode is given by

$$V_{rec} = \frac{1}{4}\gamma V_o^2 \tag{1}$$

where γ is curvature coefficient and V_o is the amplitude of the high-frequency voltage source, which depends upon the effective area of the antenna and irradiance of the solar radiation. For MIM structures, the curvature coefficient is a figure of merit. It is the measure of the non-linearity of the device. Diode with a higher value of curvature coefficient is a better rectifier. This characteristic depends on the diode material, insulator thickness, and quality of the insulator.

The cutoff frequency of the MIM diode is given by

$$f = \frac{1}{2\pi R_d C_d} \tag{2}$$

The value of the diode resistance depends mainly on the fabrication process hence the capacitance can be used to adjust the cutoff frequency of the device. The diode capacitance is given by

$$C_d = \frac{\varepsilon_r \varepsilon_o A}{d} \tag{3}$$

where ε_r is the relative permittivity of the insulating layer, ε_o is the permittivity of free space, A is the diode junction area, and d is the thickness of the oxide layer. Since the antenna resistance is usually on the order of 100 Ω, in the case of a 10 μm wavelength infrared radiation, the junction capacitance should be smaller than 0.1 fF. This implies devices with an ultra-small diode area (smaller than 50 nm × 50 nm) are required. Such small area results in a high enough cut-off frequency, making the diode capable of rectifying induced currents of the several THz frequency range.

The ultra-high speed response and a low impedance to match well with that of the antenna are the two primary requirements for the rectenna diodes. Although MIM diodes are the most mature devices for rectification at THz frequencies there is tradeoff to achieve both low resistance and low RC time constant. One possible optimization is the fabrication of MIM diodes with two insulator layers. They are called MIIM diodes [13]. Another promising THz rectifier is the geometric diode. The resistance of a geometric diode is sufficiently low to match the antenna impedance. It also does not suffer from RC constraints due to its planner structure [14].

2.2 Concentrator Lenses

The power density of the incident solar radiation on the nanoantenna can be increased by the use of concentrator lenses. Theoretically, energy conversion efficiency of such systems can be boosted by the use of spherical and cylindrical polymer micro lenses. The micro lens focuses the solar radiation on its focal point where nanoantennas can be placed. The micro lens technology in the field of photovoltaics for light trapping in order to maximize the exploitation of solar radiation has already been employed [15].

A technology for fabricating high quality micro lenses by high throughput hot embossing techniques, with size ranging from ~ 1 to hundreds of micrometers was developed in the past years [16]. With this technology, 100% area of large surfaces can be covered with lenses, which is an advantage with respect to other technologies, where typically 10−20% of the area remains un-patterned, and therefore is wasted regarding to light harvesting.

The concentration factor of the lens can be calculated by using the idealized models of spherical and cylindrical lenses and Snell's law of refraction. A situation of focusing light coming from the sun by an ideal lens on its focal point is shown in Fig. 4.

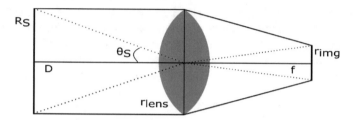

Fig. 4. A scheme of sunlight focused by a lens at its focal point

The light flux passing through the lens can be written as

$$flux = TS_1A_1 = S_2A_2 \tag{4}$$

where A_1 and A_2 are the lens and image areas, respectively, and S_1 and S_2 are respective solar irradiances. By taking the lens transmittance factor, T, as unity and replacing the area terms with the radii, the focused irradiance, S_2 can be written as

$$S_2 = S_1 \frac{r_{lens}^2}{r_{img}^2} \tag{5}$$

This can be further modified as [17]

$$S_2 = S_1 \frac{r_{lens}^2}{f^2\theta_s^2} \tag{6}$$

$$S_2 = S_1 \frac{1}{4F_{num}^2\theta_s^2} \tag{7}$$

where F_{num} is F-number of the lens. And θ_s can be estimated by the radius of the sun and its distance from the lens which could be a reasonable approximation for any object of sufficiently larger dimensions than the lens.

3   Efficiency

The conversion efficiency of the proposed energy harvesting device displayed in Fig. 2 can be evaluated by the following equation.

$$\eta = \frac{P_{out}}{P_{in}} \tag{8}$$

where P_{out} is the output DC power across the device and P_{in} is the input power collected by the nanoantenna from the solar irradiation.

To evaluate the conversion efficiency of nanoantenna-MIM diode structures combined with concentrator lenses, we have taken into account the measured rectified current on the short circuited output of the nanoantenna-MIM diode for the incident 10 μm IR radiation and the measured internal resistance of the MIM diode, in the afore-mentioned references. We have assumed, that we fill up a 1 cm^2 area with nanoantenna-MIM diode devices. A layer of micro lenses having a diameter of 50 μm is patterned over it. Under each lens, we put 20 serially connected nanoantenna-MIM diodes. In this way we can place 40000 lenses on a 1 cm^2 surface. For our theoretical calculations, we have considered measured data of Al-Al$_2$O$_3$-Pt antenna-coupled MIM diodes from Bean's thesis [18].

Infrared radiation from an average human body has been taken as input source. The irradiance level of a human body can be estimated by Planck's blackbody radiation formula [12].

$$S_\lambda = \frac{8\pi hc}{\lambda^5} \frac{1}{e^{\frac{hc}{\lambda kT}} - 1} \tag{9}$$

where S_λ is the energy per unit volume per unit wavelength radiated by a blackbody at temperature T. The spectral distribution of an average human body at 310 K is displayed in Fig. 5.

Device data used for the efficiency evaluation is summarized in Table 1. Radiation power density of 44 mW/cm^2 is calculated by integrating the human body irradiance curve over the entire wavelength range i.e. from 0 to 25 μm. By using a lens of concentration ratio 10000, the incident power density of input infrared irradiance can be increased which is estimated to be 2198 W/cm^2 by (7). Considering a linear response of Al-Al$_2$O$_3$-Pt antenna-coupled MIM diodes an output current of 0.4395 μA is estimated for this radiation. Based on these estimations, an output power of 34 mW has been calculated for 800000 diodes. This gives an efficiency of 77% for short circuit device. Considering a perfect matched load, the current will become half of the short circuit value which will reduce the efficiency by one fourth. An estimation of the efficiency with the same data but without lenses gives a value of $\sim 10^{-9}$ which is consistent with the results presented in [9].

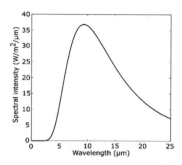

Fig. 5. Human body irradiance at 310 K

Table 1. Efficiency evaluation data

Parameter	Value	Unit
Chip size	1	cm^2
Lens size	50	µm
No. of lenses	40000	
No. of diodes	800000	
Input irradiance	0.04395	W/cm^2
Input power on chip	43.95	mW
Concentrated irradiance	2197.5	W/cm^2
Measured output current for one diode	0.4395	µA
Estimated output power of one diode	4.2494×10^{-5}	mW
Estimated output power of the chip	34	mW
Estimated efficiency	77	%

4 Conclusion

A nanoantenna-MIM diode-micro lens device has been proposed to harness energy from solar and ambient infrared radiation. The conversion efficiency of the proposed device has been investigated by utilizing the measured data of fabricated devices from the technical literature. Assuming broadband antennas and idealized lenses, an efficiency of ∼77% has been estimated.

Realization of such devices is possible by using the recent nanofabrication technologies. An antenna is an efficient device, which operates on the principle of natural resonance, therefore, it is possible to capture energy from a broad spectrum of radiation by a well-designed broadband antenna. However, optimization of MIM diode is required for efficient coupling with the antenna. Conversion of ambient infrared radiation from heated sources is also possible by these devices. They can operate at any time of the day. Another advantage is their lower cost. Since they do not require expensive semiconductor materials and the processing cost can be reduced by modern high resolution parallel pattern replication technologies, such as nanoimprint lithography. Based on its several advantages over the PV cell, the proposed device has the potential to provide a breakthrough in solar energy harvesting.

References

1. Sabaawi, A.M.A., Tsimenidis, C.C., Sharif, B.S.: Analysis and modeling of infrared solar rectennas. IEEE J. Sel. Top. Quantum Electron. **19**(3), 9000208 (2013)
2. Mescia, L., Massaro, A.: New trends in energy harvesting from earth long-wave infrared emission. Adv. Mater. Sci. Eng. **252879** (2014)
3. Bailey, R.L.: A proposed new concept for a solar-energy converter. J. Eng. Power **94**, 73–77 (1972)
4. Heiblum, M., Wang, S., Whinnery, J.R., Gustafson, T.K.: Characteristics of integrated MOM junctions at DC and at optical frequencies. IEEE J. Quantum Electron. **14**(3), 159–169 (1978)
5. Wilke, I., Herrmann, W., Kneubhl, F.K.: Integrated nanostrip dipole antennas for coherent 30 THz infrared radiation. Appl. Phys. B Laser Opt. **58**(2), 87–95 (1994)
6. Gadalla, M.N., Abdel-Rahman, M., Shamim, A.: Design, optimization and fabrication of a 28.3 THz nano-rectenna for infrared detection and rectification. Sci. Rep. **4**, 1–9 (2014)
7. Bean, J.A., Weeks, A., Member, S., Boreman, G.D., Member, S.: Performance optimization of antenna-coupled Al/AlOx/Pt tunnel diode infrared detectors. IEEE J. Quantum Electron. **47**(1), 126–135 (2011)
8. Rakos, B., Yang, H., Bean, J.A., Bernstein, G.H., Fay, P., Csurgay, I., Porod, W.: Investigation of antenna-coupled MOM diodes for infrared sensor applications. Springer Proc. Phys. **110**, 105–108 (2006)
9. Briones, E., Alda, J., González, F.J.: Conversion efficiency of broad-band rectennas for solar energy harvesting applications. Opt. Express **21**(3), A412–A418 (2013)
10. Corkish, R., Green, M.A., Puzzer, T.: Solar energy collection by antennas. Sol. Energy **73**(6), 395–401 (2002)
11. Dagenais, M., Choi, K., Yesilkoy, F., Chryssis, A.N., Peckerar, M.C.: Solar spectrum rectification using nano-antennas and tunneling diodes. In: Proceedings of SPIE, Optoelectronic Integrated Circuits II, vol. 7605, pp. 1–12 (2010)
12. Rakos, B.: Investigation of metal-oxide-metal structures for optical sensors applications. Ph. D. Thesis, University of Notre Dame, USA (2006)
13. Grover, S., Moddel, G.: Applicability of metal/insulator/metal (MIM) diodes to solar rectennas. IEEE J. Photovolt. **1**(1), 78–83 (2011)
14. Zhu, Z., Joshi, S., Grover, S., Moddel, G.: Graphene geometric diodes for terahertz rectennas. J. Phys. D Appl. Phys. **46**(18), 185101 (2013)
15. Tvingstedt, K., Dal Zilio, S., Ingans, O., Tormen, M.: Trapping light with micro lenses in thin film organic photovoltaic cells. Opt. Express **16**(26), 21608–21611 (2008)
16. Tormen, M., Carpentiero, A., Ferrari, E., Cojoc, D., Di Fabrizio, E.: Novel fabrication method for three-dimensional nanostructuring. Nanotechnology **18**, 385301 (2007)
17. Jenkins, F.A., White, H.E.: Fundamentals of Optics, 4th edn. McGraw-Hill Primis, New York (1937)
18. Bean, J.: Thermal infrared detection using antenna-coupled metal-oxide-metal diodes. Ph.D. thesis, University of Notre Dame, USA (2008)

To the Problems of Detecting Signals Passing Through a Random Phase Screen

Nugzar Kh. Gomidze[1(✉)] [ORCID], Miranda R. Khajisvili[1] [ORCID],
Izolda N. Jabnidze[1] [ORCID], Kakha A. Makharadze[2] [ORCID],
and Zebur J. Surmanidze[1] [ORCID]

[1] Batumi Shota Rustaveli State University, Batumi 6010, Georgia
gomidze@bsu.edu.ge
[2] Batumi Referal Hospital, Batumi, Georgia

Abstract. The goal of the present work is to study the regulations of the changes of the characteristics of high frequency optical signals using both quantitative and qualitative terms, on the basis of changes statistical parameters of turbulent media on the random phase screen model. There is an analytical assessment of statistical moments of laser radiation via random phase screen by numerical modeling and comparison with known experimental results. The object of the study is a random inhomogeneous atmosphere with weak turbulence, as well as optically dense turbulent media. The model of a random phase screen is discussed and the distribution of the statistical moments of scattered laser radiation is studied. The dependence of the effective size of the laser beam and the scintillation index on the correlation radius of the phase screen in the plane of the detector for a random phase screen is estimated.

Keywords: Phase screen · Statistical moments · Laser

1 Introduction

The refractive index of the media is a fluctuating value and is excited in space. The source of the ray is laser equipment because phase screen is randomly inhomogeneous media. That is why, it is obvious that the inclination of the ray from the straight direction takes place. The propagation of a ray in randomly inhomogeneous media should be considered as a stochastic process and the Einstein-Fokker-Kolmogorov equation [1] for angular separation of rays can be used. The Einstein-Fokker-Kolmogorov equation makes it possible to calculate the angular and linear inclination from the source direction.

In fundamental point of view, the processes of natural fluctuation are very interesting, which take place on phase screen (drinking water, sea water, liquid crystal etc.). Molecular scattering of light in from phase screen is connected to the thermal fluctuation processes, which represent the function of reflection indicator, density, pressure, temperature, entropy and molar concentration [2]. According to Onsager's hypothesis, the fluctuation processes in water can be described with the macroscopic hydrodynamic rules. And fluctuation attenuation is described with time auto-correlative function, the Fourier-transform of which gives the analytical expression of the intensity of the optical

© Springer Nature Switzerland AG 2019
G. Laukaitis (Ed.): INTER-ACADEMIA 2018, LNNS 53, pp. 177–184, 2019.
https://doi.org/10.1007/978-3-319-99834-3_24

spectrum, according to Wiener-Khintchin theorem. Of course, the chaos of fluctuation processes in water makes us consider the diffusion processes as well (Einstein-Stokes formulae), which are finally reflected in analytical expression of the spectrum intensity.

It should be mentioned that in the superhigh frequency range the analogical methods of signal processing are well known [3]. But since the laser sources of the spectrum have been invented the analogical methods were widely applied in spectroscopy as well. As for the numerical methods of signal processing, which are discussed in the present paper, are one of the actual problems.

Numerical methods were applied from the photon counting experiments which were made in different laser source for studying statistical features. On the basis, of numerical methods of signal processing have worked out highly effective digital, rapid auto-correlator, which was worked in real mode [4] and which enabled to provide dimensions in wide frequency range 1–10^8 GHz. The purpose of the auto-correlator was to widen the frequency range lower than optical MHz. this problem was solved by Farbi-Perot's interferometer [5].

On the basis of statistical theory, within the limits of the project FR/152/9-240/14 was developed numerical correlator, which generated of pseudo-random impulses so that the transferred signal spectrum wasn't dependent on the impulse length (on the number of impulses in the signal).

Widening the frequency spectrum of the carrier signal was happen via angular modulation (PM – Phase Modulated, FM – Frequency Modulated) [6], in particular, via modem. Between the modulator and demodulator was a perfect synchronization – coherent mode. Superhigh frequency signal is transmitted to the phase modulator from the digital generator, which manages the impulses with random sequence. At the entrance of the modulator, the signal will have an image of continues noise-shaped spectrum. This signal is joined to a wide frequency striped external signal – "noise" of non-Gaussian statistics. The process "modulation – demodulation" can be represented as random phase screen simulative model [7, 8].

The statistical property of propagation of a Gaussian beam through a phase screen with arbitrary thickness is investigated in the work [9].

In the present paper is investigated Influence of the detector surface on the intensity spectrum of the fluctuation and Doppler's spectrum for the Gaussian light on the basis wave theory Rytov and thin-phase screen model Andrews [10]. The effective normal radius of beam and the effective scintillation index for the correlation radius of the fluctuations of a random phase screen are estimated.

2   Problem Statement

Let's say, on the $z = 0$ plane the primary u_0 field statistics are given, i.e. its moments (coherence function) are given. It is necessary to find out how these functions are changing away from the $z = 0$ plane, if on the way the field is transformed (for example, the wave is going through a diaphragm, lens or other).

Formally this task is solved easily: If it is known how the determined (totally coherent) wave changes, then it is enough to make an ensemble averaging of the determined solution of u_0 field. But this method, as a rule, reduced to difficultly calculating integrals. For example, when calculating the relative fluctuations of intensities $\sigma = \langle I^2 - \langle I^2 \rangle \rangle / \langle I^2 \rangle$ (the same as the "scintillation index") it is necessary to calculate the eight-fold integral, which is practically impossible to be calculated even in the case of simplified model. The simplified marginal model implies that a flat wave e^{ikz} falls on the layer, directly on the other side of the screen $u_0 = e^{ikz + \Psi}$, where $\Psi(x, y)$ is the random phase. Knowledge of phase statistics on the screen determines the field statistics on $z = 0$ flatness. The "system" which "transforms" the field out of the screen is just a free space in the given case. As a result of diffraction, the wave, which passes through the chaotic phase screen, is experiencing fluctuations. However, the intensity of the phase screen is permanent (Fig. 1a).

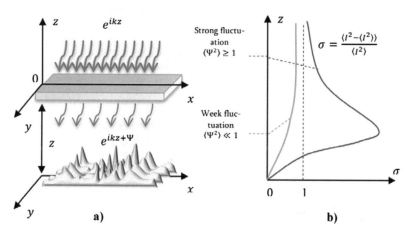

Fig. 1. (a) Distribution intensity of wave threw random phase screen. (b) Dependence scintillation index on distance from random phase screen for week $\langle \Psi^2 \rangle \ll 1$ and strong $\langle \Psi^2 \rangle \gg 1$ fluctuation of phase

Fortunately, for the well-established objectives in physical terms, approximate calculation methods for complex and multiple integrations can be found. In the private case, the "scintillation index" σ for the phase screen can be calculated during the weak phase fluctuations $\langle \Psi^2 \rangle \ll 1$ of the phase. In case of strong fluctuations $\langle \Psi^2 \rangle \geq 1$ - to calculate σ, on small z distances, can be applied an excitation method (considering that the fluctuations of the intensity are small beyond the screen) and for the longer distances can be applied a method of normalization of the field. The normal law of probability distribution implies that in the long z distances in the point of observation, many non-correlated waves take place from different parts of the screen. From these districts in the middle district - in the so called, focus area, can be found the asymptotic of the field during the presence of large fluctuations in the phase. i.e. when $\langle \Psi^2 \rangle \gg 1$. As a result had got $\sigma(z)$ curves the qualitative image of which is shown at Fig. 1b).

Here is considered a case when the laser beam is perpendicular to the phase screen. The phase screen is of a small thickness, scattered radiation is registered in the distant zone through the photo detector. The photo detector axis makes θ angle of the laser beam (Fig. 2).

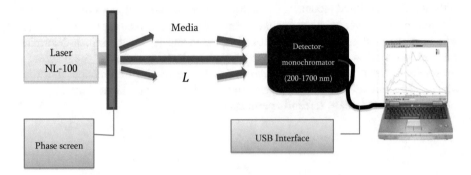

Fig. 2. An experimental device scheme for studying partially coherent laser beam in the free space or in the turbulent media. $L > > L_f$, L_f- is the distance from diffuser to the detector, the L_f - the focal length of the optical system

The problem of evaluating the characteristics of a laser beam during the passage the random phase screen can be brought to an adapted selected model to the task, so that the estimating characteristics of the beam should be more sensitive to the characteristics of the random phase screen.

3   Discussion and Results

The optical signal features are discussed on the example of an ideal, stable, single-mode laser source:

$$s(t) = s_0 \cos(\omega_0 t + \Psi)$$

In the process of scattering the modulation takes place, the random statistics of the phase screen causes the creation of Doppler effect - displacement towards the optical frequency. Phase fluctuations causes the broadening of radiation spectrum to the GHz towards the optical frequency ω. The signal power spectrum or auto-correlation function cannot provide complete information about the process of scattering, but these values are subject to experimental measurement, therefore, the theoretical and experimental measurement of these parameters may become the basis for creating an adequate theoretical model. From this perspective, the first and second order statistical moments are presented as experimental measurable physical values on the basis of the correlation functions of the scattered signal and intensity:

$$g^{(1)}(\tau) = \frac{\langle s^+(t)s^-(t)\rangle}{\langle I^2\rangle}, \quad g^{(2)}(\tau) = \frac{\langle I(0)I(\tau)\rangle}{\langle I^2\rangle}$$

where, $I(t) = s^+(t)s^-(t)$ - represents a magnitude that is recorded by a detector. In the quantum-mechanical point of view, the detector, which works on photon registration (through photo multiplier and photodiode), records this value. $s^+(t)$ and $s^-(t)$ represents the signal components that correspond to positive and negative frequencies and are determined by the Fourier series:

$$s(t) = \sum_{\omega>0} a_\omega^* e^{-i\omega t} + \sum_{\omega\geq0} a_\omega e^{-i\omega t} = s^+(t) + s^-(t), \quad a_\omega = \frac{1}{T}\int_{-T/2}^{T/2} s(t)e^{-i\omega t} = a_{-\omega}^*.$$

It is noteworthy that from $g^{(1)}(\tau)$ and $g^{(2)}(\tau)$ statistical moments only the second order statistical moment $g^{(2)}(\tau)$ is to be measured on the detector. That's why, the matter of finding the connection between $g^{(2)}(\tau)$ and $g^{(1)}(\tau)$ statistical moments is an urgent issue. This connection was found for statistically independent random variables on the basis of probability theory and was brought to the expression that is a well-known Siegert's ratio and which allows the first-range spectral characteristics to be calculated by the simplest spectral characteristics of the second order:

$$g^{(2)}(\tau) = \langle I(t)I(t+\tau)\rangle = \langle I^2\rangle\left(1 + \left|g^{(1)}(\tau)\right|^2\right)$$

Similar calculations were conducted for a specific task when the scattered field:

$$s^+(t) = s_0\exp[-i(\omega_0 t + \varphi_0(t))] + f(t)s_1\exp[-i(\omega_1 t + \varphi_1(t))]$$

and primary source field were radiated from the same point of space and the intensity correlation function was calculated, which is the function of $g^{(1)}(\tau)$ ꝏ $g^{(2)}(\tau)$ moments:

$$g^{(1)}(\tau) = \langle f(0)f(\tau)\rangle, g^{(2)}(\tau) = \langle f^2(0)f^2(\tau)\rangle$$

It was shown here that making Doppler spectrum dominant in the expression of correlation function of intensity, increases the significance of the signal/noise ratio on the detector's output in connection with the fluctuation spectrum of intensity.

In the work [4] have evaluated the impact of the detector area on the fluctuation intensity spectrum and on the Doppler spectrum for Gaussian light:

$$f(S) = \sum_{R=0}^{\infty}\left[\frac{(2R+2)!}{[(R+1)!]^2(R+2)!}\right]^2(-1)^R\left(\frac{1}{2}\chi R'\right)^{2R}$$

$$f_D(S) = \frac{4}{\pi R'^4} \int_0^{R'} \int_0^{R'} r_1 r_2 dr_1 dr_2 \int_0^{2\pi} d\varphi \frac{J_1(\chi\sqrt{r_1^2 + r_2^2 - 2r_1 r_2 \cos \varphi})}{\chi\sqrt{r_1^2 + r_2^2 - 2r_1 r_2 \cos \varphi}}$$

where $\chi = k_0 R/z$, R - is the radius of circular source, R' - is the radius of the surface of photocathode of the detector. Influence of the detector surface on the intensity spectrum of the fluctuation $f(S)$ and on the Doppler's spectrum $f_D(S)$ for the Gaussian light is given on Fig. 3.

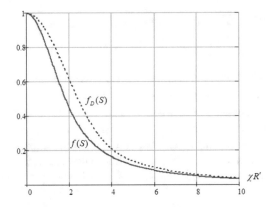

Fig. 3. Influence of the detector surface on the intensity spectrum of the fluctuation $f(S)$ (red line curve) and on the Doppler's spectrum $f_D(S)$ (blue dot curve) for the Gaussian light

The positive-frequency part of the electric field of the light wave that passes the phase screen can be written by the following expression:

$$\xi^+(\vec{r}, 0; t) = E_0 \exp\left\{ i[\phi(\vec{r}, t) - \omega_0 t] - \frac{r^2}{W_0^2} \right\},$$

where $\phi(\vec{r}, t)$ is the random phase shift dependent on the spatial coordinates and are conditioned by the phase screen. The phase screen is located in the $z = 0$ plane. W_0- is the width of the Gaussian distribution of the intensity of flow and it is characterized to the size of the illuminated area.

Based on the first and second orders of the laser radiation scattered from the random phase screen, can be received:

$$<I(\theta; t)> = \pi^2 W_0^2 |E_0|^2 \int_0^\infty dr\, r J_0(kr \sin \theta) \exp\left\{ [-\bar{\phi}^2(1 - \rho(r))] - \frac{r^2}{W_0^2} \right\}.$$

$$\frac{<I^2>}{<I>^2} = 2 - \frac{2l^2}{W_0^2} + \frac{l^2\bar{\phi}^2}{4W_0^2} \exp\left[\frac{k^2 l^2 \sin^2 \theta}{4\bar{\phi}^2} \right].$$

when $l/W_0 \rightarrow 0$, $<I^2>/<I>^2 = 2$, that is characterized to a Gaussian statistics. When $l \ll W_0$ the distribution is significantly different from Gaussian and the existence of $\bar{\varphi}^2$ circumstance, which is quite large in terms of access. Finally, when $l \ll W_0$ the second order moment may be more than 2 if $\bar{\phi}^2$ is a big enough value. Moreover, this effect will increase even more by increasing θ.

The dependence of the effective normal radius $W_{1,d}/W_0$ ($W_0 = 2.5$ cm) of the laser beam on the correlation radius l_c via diffuser is shown on the Fig. 4(a). Shows an effective scintillation index as the function of the correlation radius (red line). The scintillation index, which is in the work of is presented in the work [11] with green line.

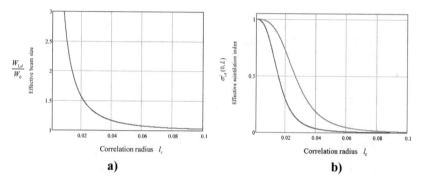

a)   b)

Fig. 4. (a) Dependence of the effective normal radius of beam W_{1d}/W_0 vs correlation radius l_c fluctuations of intensity phase screen. (b) Effective scintillation index vs correlation radius l_c for the parameters: $W_0 = 2.5$ cm, $\lambda = 0.5$ mkm, $L = 500$ m

4 Conclusion

The laser beam size, intensity, correlation radius and the expressions of the angle of fluctuations are obtained in the plane of the detector. It is shown that the laser beam size, the correlation radius and the average intensity of the radiation coincide with the classical results. However, the results obtained in the turbulence of the atmosphere somewhat differ from the classical results. The scintillation index is obtained in the free space of laser beam, as well as in the weak turbulence atmosphere. In case the strong diffuser, the model shows that the index is close to 1. When correlation radius of the intensity fluctuations of the phase screen is increasing the effective radius of the normal beam passing through phase screen decreases and goes to 1, which corresponds to the case that the beam expansion is insignificant at large correlation radius. The dependence of the effective scintillation index on the correlation radius gives the same result as in the work [11]. But, in our case, the scintillation index is more sensitive depending on the correlation radius of the intensity fluctuations, and therefore, the proposed model is more sensitive in estimating the fluctuation processes of the phase screen.

Acknowledgements. Work it was spent within the limits of the scientific projects: "Quantitative analysis of fluorescence characteristics of optically solid, random phase screen and spectral analysis of the statistical moments of the correlation function of the intensity of scattered laser radiation" (FR/152/9-240/14 scientific supervisor Prof. Nugzar Gomidze), financed by SRNSF (Shota Rustaveli National Scientific Found, Georgia).

References

1. Libermann, L.: Effect of temperature inhomogeneities in the ocean on the propagation of sound. J. Acoust. Soc. Am. **5**, 23 (1951)
2. Ritov, S.M., Kravtsov, Y.A., Tatarsri, V.I: Introduction in statistical radiophysics. V.2, Random fields. M., "Nayka" (1978)
3. Forrester, A.T., Gudmundsen, R.A., Johnson, P.O.: Phys. Rev. **99**, 1691 (1955)
4. Pike, E.R., Jakeman, E.: Photon Statistics and Photon Correlation Spectroscopy. Academic Press, Cambridge (1973)
5. Jekemen, E.: Photon Correlation (1974)
6. Gomidze, N., Khajishvili, M., Makharadze, K., Jabnidze, I.: Some features of radio-spectral diagnostics of random media via PM and PRM oscillations. J. Appl. Mech. Mater. **420**, 305–310 (2013)
7. Gomidze, N.Kh., Khajishvili, M.R., Makharadze, K.A., Jabnidze, I.N., Surmanidze, Z.J.: About statistical moments of scattered laser radiation from random phase screen. Int. J. Emerg. Technol. Adv. Eng. **6**(4), 237–245 (2016). ISSN 2250-2459
8. Khajishvili, M.R., Gomidze, N.Kh., Makharadze, K.A., Jabnidze, I.N., Chikhladze, M.: Evalution bandwidth of optical signal via statistical moments of phase screen. Int. J. Polytech. **21**(1), 123–128 (2018). https://doi.org/10.2339/politeknik.379641
9. Tian, Y., Guo, J., Wang, R., Wang, T.: Mathematical model analysis of gausian beam propagation through an arbitrary thickness random phase screen. Opt Exp. **19**, 18216 (2011)
10. Andrews, L.C., Philips, R.C.: Laser Beam propagation through random media. SPIE Optical Engineering Press, Bellingham (1998)
11. Korotkova, O., Andrews, L.C.: Speckle propagation through atmospheric turbulence: effects of partial coherence of the target (SPIE pr.) (2002)

Robotics, Measurement, Identification, and Control

Vehicle Detection Using Aerial Images in Disaster Situations

Ayane Makiuchi[✉] and Hitoshi Saji

Graduate School of Integrated Science and Technology, Shizuoka University,
3-5-1, Johoku, Naka-ku, Hamamatsu, Shizuoka 432-8011, Japan
makiuchi.ayane.17@shizuoka.ac.jp

Abstract. In disaster situations, it is necessary to rapidly determine the locations of traffic jams and abandoned vehicles to find traffic routes so that rescue activities can be carried out efficiently. However, it takes time to recognize vehicles and estimate their positions. The purpose of our study is to rapidly detect vehicles and their positions in the case of disasters. We propose a method of vehicle recognition and position estimation using an aerial image taken from a helicopter and a road map. Although such images can be taken locally, occlusion occurs when buildings are reflected on the road. For vehicle recognition, we use shadow correction, asphalt removal by machine learning, and shape analysis. In addition, we remove buildings to solve the problem of occlusion. First, we adjust the color of the aerial image by shadow correction. Then, we remove areas of asphalt and buildings on the road, and we extract vehicle areas by using their shape features. To estimate the positions of vehicles, we project the road map on the aerial image by a projective transformation. We extract the road area from the aerial image by the projection before vehicle recognition, thus increasing the efficiency of the process. Using our method, we successfully detected most vehicles with owing to their different colors from the asphalt. Furthermore, we marked the positions of the vehicles on the road map. We thus demonstrated the possibility of the rapid detection of vehicles from aerial images in disaster situations.

Keywords: Vehicle detection · Aerial image · Earthquake

1 Introduction

Massive earthquakes occur frequently in some parts of the world, causing serious damage. After an earthquake, any delay in rescue activities may result in the loss of human life. The analysis of aerial images is one of the methods of collecting traffic information. The advantage of this method is that information can be obtained over a large area regardless of the state on the ground. In recent years, various methods have been established using high-resolution cameras or UAV [1], and many of which focus on overhead images [2].

However, such limited methods are unsuitable for disaster situations. This is because of the likelihood of blocked roads and difficulties in procuring specific devices. In fact, the videos and images released by the Geographical Survey Institute immediately after

© Springer Nature Switzerland AG 2019
G. Laukaitis (Ed.): INTER-ACADEMIA 2018, LNNS 53, pp. 187–194, 2019.
https://doi.org/10.1007/978-3-319-99834-3_25

the earthquakes are oblique side views captured from helicopters with handheld cameras and drones. Under such circumstances, it is impossible to use conventional methods employing images which are high-resolution or overhead. Immediacy should be given the highest priority in traffic analysis after a strong earthquake, and a method that is not limited by conditions on the ground is required. Therefore, we propose a method for detecting vehicles from oblique side-view aerial images.

In oblique side-view aerial images, vehicles are captured from various angles and are not homogeneous. Therefore, we cannot detect vehicles by simple matching using shape features. There are methods for detecting objects that employ 3D models [3], but they are unsuitable after a strong earthquake because they require sensor cameras and take time. Furthermore, in oblique side-view aerial images, buildings are reflected in areas of road.

On the basis of these issues, we do not focus on vehicles that they are not homogeneous in this study. After removing homogeneous asphalt areas, processing is performed efficiently on low-resolution images. Furthermore, by clearly distinguishing between buildings superimposed on the road and the vehicles, vehicles can be distinguished in oblique side-view images.

2 Proposed Method

This method is divided into three steps: road extraction, asphalt and building removal, and the output of results.

In the road extraction step, we use a projective transformation to map a road map to an oblique side-view aerial image. In the asphalt and building removal, we use machine learning, shadow correction, and edge extraction to eliminate the asphalt areas, and we focus on the periphery of the road to exclude the building areas. This step is substantially vehicle detection step because it is possible to leave only the vehicles remain after this step. Before outputting the result, we perform noise removal and small-region integration on the vehicle discrimination results to make them visually easy to understand.

2.1 Road Extraction

Since the aim of this study is to collect traffic information, we limit vehicle detection to road areas. The advance extraction of roads from the aerial image accelerates subsequent processing.

We extract road areas by masking the map image on the aerial image. However, we cannot use simple masking for extraction because of the obliqueness of the aerial image. Therefore, we use a projective transformation and deform the road map to match the aerial image. The conversion parameters are calculated by specifying four corresponding points in the aerial image and the road image.

2.2   Preprocessing of Asphalt and Building Extraction

We perform region-based determination of asphalt and buildings to exclude them correctly. Before the determination, we perform smoothing the aerial image and divide it into regions.

For the smoothing, we use edge-preserving regularization [4]. This method smooths images at a high speed while preserving vehicle edges, so it is suitable for our study. For region division, we use the mean shift [5].

2.3   Asphalt Extraction

A schematic diagram of the proposed method is shown in Fig. 1. In this process, we use shadow correction, edge features, and machine learning to comprehensively determine whether each region divided in the preprocessing is asphalt.

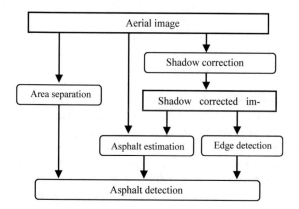

Fig. 1. A schematic diagram of the proposed method

2.3.1   Shadow Correction

Since the regions covered with shadows of buildings change in color, it is difficult to distinguish between asphalt and vehicles. In addition, the edges of shadows may be extracted as those of objects in the later processing. To solve these problems, we correct the shadows in the aerial image.

We use the method of S. Murali et al. for shadow correction [6]. This method determines whether a pixel is in shadow from the value in the LAB color space. The values of the pixels determined to be in shadows are corrected as follows.

$$\begin{cases} L+ \ = 28 \\ A- \ = 3 \\ B- \ = 5 \end{cases} \tag{1}$$

2.3.2   Edge Features

Asphalt areas don not have edges because of the uniformity of the asphalt color. In contrast, vehicle areas have many edges that define their body and windows. To use these effectively, we extract edge features from the shadow-corrected aerial image.

However, the white lines on the roads may be mistaken for vehicle edges. To blur the white lines, we use non-local means algorithm [7] before edge extraction. This method eliminates noise in the image while retaining edges, so it reduces the edge strength of white lines while retaining the edges of vehicles. After denoising, we extract edges as thin lines by the method of Canny [8].

2.3.3   Estimation by Machine Learning

In addition to edge features, we use machine learning to estimate whether each pixel is asphalt on the basis of its value. Specifically, we use the EM algorithm to estimate of a mixed Gaussian distribution [9]. The learning data are oblique aerial images taken from a helicopter above Kumamoto after the 2016 Kumamoto Earthquake. The distribution generated by the learning classifies each pixel into asphalt or not.

This classification is performed on the images before and after the shadow correction. Since the pixels in shadows are easily discriminated as erroneous, the classification of these two images increases the reliability of results.

2.3.4   Judgement Condition for Asphalt

We determine areas of asphalt from the edge features and by estimation based on machine learning. For the determination, total number of pixels in the area, edge pixels, and pixels judged to be asphalt by machine learning are counted. The total number of pixels in each region, the edge pixels, pixels judged to be asphalt before shadow correction, and pixels judged to be asphalt in the image after shadow correction and are respectively defined as NoP, $edge$, ML, and ML_{shadow}, and each area is judged to be asphalt if it satisfies the following condition.

$$\frac{(NoP - edge) * 0.6 + ML * 0.2 + ML_{shadow} * 0.2}{NoP} \geq 0.6$$

2.4   Building Extraction

Building extraction is carried out in the areas remaining after the extraction of asphalt. Buildings are eliminated from the areas in contact with the road edges. In this process, we extract areas touching the road edges as buildings. This is because a building includes areas on both sides of a road edge when it is reflected in a road.

2.5   Output of Processing

From the above process, we can detect areas of candidate vehicles in the aerial image. However, in these areas, there may be small areas remaining after the other areas have been extracted, or a single vehicle may be divided into multiple areas. Therefore, we remove the areas that are too small to be vehicles and amalgamate the divided vehicle areas.

Expansion and Contraction Treatments. We use expansion and contraction treatments to refine the results. In this process, noise removal and filling are performed on a binary image. This process removes small areas and area integrates multiple areas.

3 Experiment

We apply the proposed method to actual oblique aerial images and confirm its effectiveness.

3.1 Experimental Environment

In this experiment, we use aerial images and map images published by Geographical Survey Institute. Which were taken from a helicopter above Kumamoto prefecture after the 2016 Kumamoto Earthquake. The size of each original 24-bit color aerial image is 983×655. However, we remove the bottom of the image containing information such as the shooting time, resulting in an image size of 983×600. The size of the map image was 820×592, and those around the location of the aerial image are used. In this experiment, we use two pairs of aerial and map images and name them Pair 1 and pair 2.

3.2 Accuracy Evaluation

We evaluate the accuracy of the above two pairs of results. It is assumed that detection is successful if 80% or more of the detected areas overlap with vehicles. We evaluate the accuracy by finding the f value and the rate of correct detections.

3.3 Experimental Results

The input images of Pair 1 used are shown below. Figure 2 shows one of the aerial images and Fig. 3 shows the road mask image created from the map image. The mask image is a binary image with road regions colored in white and non-road regions colored in black. Figures 4 and 5 shows results of vehicle detection of Pair 1 and Pair 2. The regions detected as vehicles are colored in red. Table 1 shows the accuracy evaluation of the both pairs.

Fig. 2. Aerial image (Pair 1)

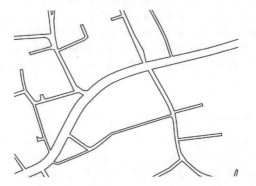

Fig. 3. Map image of (Pair 1)

Fig. 4. The result of the vehicle detection (Pair 1)

Fig. 5. The result of vehicle detection (Pair 2)

Table 1. Accuracy evaluation

	Precision	Recall
Pair 1	0.72	0.75
Pair 2	0.93	0.67
Average	0.83	0.71

3.4 Discussion of Experiment Results

Among the final results in Figs. 4 and 5, many of the vehicles that were detected correctly had bright colors such as white. This is because they have many edges owing to the large difference in color between their bodies and their window and the surrounding asphalt. In contrast, the vehicles with similar colors to asphalt, such as gray and black vehicles, had low detection accuracy.

We were able to successfully eliminate white lines and buildings, which may be easily confused with cars. This is because the smoothing before edge extraction and the exclusion of buildings were effective. In contrast, many of the erroneous detections were due to telegraph poles. Since telegraph poles are reflected across the road similarly to buildings, they should be excluded in when excluding buildings. However, since telegraph poles are very thin and the image was finely divided at the time of region division, erroneous discrimination occurred.

4 Conclusion

In this paper, we propose a method of distinguishing vehicles from images taken obliquely after a strong earthquake. Effective use at more earthquake sites is expected by a traffic analysis method that does not use special equipment such as high-performance cameras. From now on, we will attempt to improve the discrimination accuracy of vehicles and a system that can detect blocked roads.

Acknowledgements. This work was supported by JSPS KAKENHI Grant Numbers JP18K04657.

References

1. Moranduzzo, T., Melgani, F.: Automatic car counting method for unmanned aerial vehicle images. IEEE Trans. Geosci. Remote Sens. **52**(3), 1635–1647 (2014)
2. Tang, T., Zhou, S., Deng, Z., Lei, L., Zou, H.: Vehicle detection in aerial images based on region convolutional neural networks and hard negative example mining. Sensors **17**(2), 336 (2017)
3. Tani, I., Yokoi, A., Saji, H.: Vehicle extraction using geometrical constraint. J. Inst. Image Electron. Eng. Jpn. **43**(4), 579–587 (2014)
4. Charbonnier, P., Blanc-Fraud, L., Aubert, G., Barlaud, M.: Deterministic edge-preserving regularization in computed imaging. IEEE Trans. Image Process. **6**(2), 298–311 (1997)
5. Comaniciu, D., Meer, P.: Mean shift: a robust approach toward feature space analysis. IEEE Trans. Pattern Anal. Mach. Intell. **5**, 603–619 (2002)
6. Murali, S., Govindan, V.K.: Shadow detection and removal from a single image using LAB color space. CIT **13**(1), 95–103 (2013)
7. Buades, A., Coll, B., Morel, J.-M.: A non local algorithm for image denoising. Comput. Vis. Pattern Recognit. **2**, 60–65 (2005)
8. Canny, J.: A computational approach to edge detection. IEEE Trans. Pattern Anal. Mach. Intell. **8**(6), 679–698 (1986)
9. Bahl, L.R., Jelinek, F., Jelinek, R.L.: A maximum likelihood approach to continuous speech recognition. IEEE Trans. Pattern Anal. Mach. Intell **PAMI-5**, 179–190 (1983)

On the Role of Shaped-Noise Visibility for Post-Compression Image Enhancement

Kousuke Kawai, Damon M. Chandler$^{(\boxtimes)}$, and Gousuke Ohashi

Department of Electrical and Electronic Engineering, Shizuoka University,
Hamamatsu, Shizuoka 432-8561, Japan
chandler.damon.michael@shizuoka.ac.jp
http://vision.eng.shizuoka.ac.jp

Abstract. Digital images undergo irreversible compression to enable transmission, a process which induces visible and often annoying distortions. Post-compression enhancement algorithms attempt reduce the visual impacts of these distortions. One such enhancement technique, designed specifically for textures, employs perceptually shaped noise patterns. Although this technique is quite effective, the contrasts of the patterns must be properly scaled, thus necessitating an algorithm that can automatically predict these scaling factors. To help enable such a prediction, we investigated whether patterns which have higher contrast detection thresholds (CTs) [i.e., patterns which require more contrast to be just-visible] also require higher contrast scaling factors (CSs) for proper enhancement. We measured CTs in the current study and compared them with CSs measured in our previous study. Our results support the hypothesis that low CT implies low CS, and vice-versa, but only for those patterns which could markedly improve the visual quality.

Keywords: Enhancement · Restoration · Compression
Visual perception

1 Introduction

The widespread use of digital multimedia and network-capable devices has to today made the internet the most prevalent means of delivering digital images and videos. These days, rarely does one simply take a photo or record a video for local archival and consumption; rather, these data are often stored online and shared with others via social networking sites. Yet, despite technical advances in internet speeds, even today's fastest networks cannot handle the enormous bandwidth required for uncompressed (raw) images/videos. Instead, the vast majority of images/videos are encoded in a more compact format via the use lossy compression such as JPEG, JPEG-2000, or HEVC. The tradeoff, however, is that the more compression one applies, the larger are the induced digital artifacts (distortions), which can severely degrade the visual quality.

Over the last several decades, research in the field of lossy image/video compression has given rise to moderate improvements in coding efficiency. However,

© Springer Nature Switzerland AG 2019
G. Laukaitis (Ed.): INTER-ACADEMIA 2018, LNNS 53, pp. 195–203, 2019.
https://doi.org/10.1007/978-3-319-99834-3_26

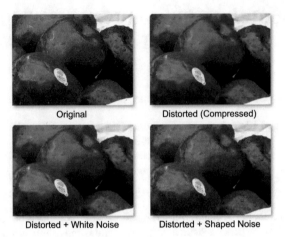

Original   Distorted (Compressed)

Distorted + White Noise   Distorted + Shaped Noise

Fig. 1. Perceptually shaped noise can improve the quality of a compressed image to a greater extent than white noise. Here, the texture of the skin on the apples, which is largely lost in the compressed image (upper-right image), has been corrected via the addition of white noise (lower-left) and our proposed technique (lower-right).

the onus of producing a high-quality image/video need not be placed entirely on the compression scheme. Digital correction/enhancement at the receiving end (i.e., after decompression) has the potential to mitigate the impacts of the distortions. Focusing specifically on images, previous research on enhancement and/or restoration of compressed images has focused on two main areas: (1) reducing the visibility of blocking artifacts, which most often appear as artificial streaks and stripes in smooth/blank areas (e.g., [8]); and (2) reducing the appearance of aliasing artifacts whereby edges and object boundaries appear artificially jagged (e.g., [2]). Very little work has focused on reducing distortions in textured regions. Indeed, when textures are compressed, fine details (higher frequencies) are lost, thus giving rise to the appearance of blurring. When this occurs, owing to the stochastic nature of textures, there is usually insufficient information remaining in the middle and lower frequency bands to facilitate even minor improvements.

We have recently proposed an alternative enhancement technique which makes use of what we call "perceptually shaped noise" to improve the appearances of compressed texture regions [11]. Our technique is based, in part, on earlier related works which have investigated the efficacy of additive white noise for improving the perceived sharpness of blurred images [3–5,9] and compressed images [1]. In [11], we demonstrated that rather than using white noise, improved visual quality could be achieved in most cases by using, not white noise, but noise that has been shaped in terms of statistical and spatial-frequency properties to better perceptually match each texture region (see Fig. 1).

Yet, in order for the proposed enhancement method to be effective, the shaped-noise patterns must be applied in the proper amounts. If the contrast of the pattern is set too high, the result appears artificial. Conversely, if the contrast of the pattern is set too low, the improvements cannot be seen. Thus, there is a need to automatically predict the optimal contrast scaling factor on a per-texture basis. We hypothesize that one determining factor is the *visibility* of the pattern when viewed within the distorted texture. Specifically, because the shaped-noise patterns are texture-specific, the patterns themselves have different levels of visibilities for a given fixed contrast—some patterns are quite readily visible, whereas others are more visually subtle. Furthermore these visibilities can change when the patterns are added to their respective distorted textures.

Thus, in this paper, to test this hypothesis, we conducted a psychophysical experiment to measure contrast thresholds for detecting the restorative patterns shown within their respective blurred textures. (The contrast detection threshold for a given pattern is defined as the minimum contrast required for the pattern to be visible.) Our specific aim was to investigate if there exists a relationship between the restorative patterns' optimal contrast scaling factors and the restorative patterns' contrast detection thresholds. Here, we describe the study and present our analysis of the results.

2   Methods

2.1   Apparatus and Subjects

Stimuli were displayed on a Display++ 32-inch LCD monitor (Cambridge Research Systems, Cambridge, UK) at a resolution of 1920 × 1080 pixels. The display had a luminance gamma of 2.2, with minimum, maximum, and mean luminances of, respectively, 0.6, 130.0, and 42.8 cd/m^2. Stimuli were viewed binocularly through natural pupils in a darkened room at a distance of ≈46 cm.

One female and four males served as subjects in the study. The first author served as one of these subjects. The subjects were all college students with ages in the range of 21–25 years. All subjects had prior experience viewing distorted images, but only the first author was informed of the purpose of the study. All subjects had self-reported normal or corrected-to-normal vision.

2.2   Stimuli

Source and Distorted Textures. The same 15 images used in [11] served as the original source images from which the textures were cropped and synthesized. The 15 images were selected from the various categories of the McGill Image Database [6], then resized via bicubic resizing with antialiasing to 768 × 576 or 576 × 768 pixels, and then converted to grayscale. The images were then hand-segmented to isolate different textures, resulting in 1–7 possibly non-contiguous textures regions per image, resulting in a grand total of 50 texture regions. A 256 × 256 region was then cropped from each source texture region. We refer to these 50 crops as the *source textures*.

Distorted versions of the original source images were the same as in [11], which were created by using HEVC single-frame compression (BPG format) with quantization scaling factors ranging from 33 to 40 to obtain roughly medium-low-quality images as judged by the first author (bit-rates ranging from 0.03 to 0.36 bpp). The same 256×256 region selected for each corresponding source texture was then cropped from each of the 50 distorted texture regions. We refer to these 50 crops as the *distorted textures*.

Fig. 2. Each horizontally arranged group of three images contains, from left to right, the distorted texture, the restorative pattern, and the enhanced distorted texture.

Shaped-Noise Restorative Patterns. The shaped-noise restorative patterns were 256×256 crops taken from the 50 patterns used in [11], which were generated via a parametric texture-synthesis algorithm [7]. This algorithm takes as input a sample of a source texture, and then generates a sample of noise whose various statistical features are then iteratively matched to the source texture. The result is a statistically shaped noise pattern that resembles the source texture to various degrees. As in [11], further perceptual shaping of the synthesized textures was performed by synthesizing only the medium- and high-frequency components, and by adjusting the local mean of each pattern to ensure constant contrast when applied to each distorted texture. We refer to these 50 synthesized patterns as the *restorative patterns*. (See [11] for further details of the source and distorted textures and the restorative patterns.)

Additive Enhancement Process. Let S_i and D_i denote the ith source and distorted textures, where $i \in [1, 50]$. Let P_i denote the restorative pattern generated from S_i. The enhancement of the distorted texture D_i by using P_i was performed as follows:

$$\tilde{D}_i = D_i + c_i P_i \tag{1}$$

where $\tilde{D}_i$ denotes the distorted texture that has been enhanced, and where c_i denotes a scaling factor that controls the contrast of the restorative pattern. In [11], c_i was explicitly adjusted by observers to maximize the visual quality. In the current study, c_i was adjusted automatically and adaptively via the psychophysical testing software (described next) to determine the minimum contrast required for each restorative pattern to be visible (i.e., the contrast detection threshold).

Figure 2 shows the 50 distorted textures (D_i), the restorative patterns (P_i), and the enhanced versions of the distorted textures ($\tilde{D}_i$). To promote visibility in this figure, all restorative patterns have the same contrast scaling of $c_i = 4$. Observe that this constant $c_i = 4$ is suitable for some textures, but too high for others, and in a few cases too low.

2.3   Procedures

Contrast detection thresholds were measured by using a spatial three-alternative forced-choice procedure. On each trial, observers concurrently viewed three adjacent images placed upon a uniform 20 cd/m^2 background. Two of the images were of the distorted texture alone, and the other image additionally contained one of the previously described restorative patterns. Subjects were asked to indicate which of the three images contained the restorative pattern. The contrast of the restorative pattern was controlled via an adaptive staircase procedure (QUEST [10]) in which the next test contrast was determined based on the subject's response history. If the subject correctly chose the image containing the restorative pattern, then the next test contrast would be reduced (restorative pattern made more difficult to see). Conversely, if the subject did not correctly choose the image containing the restorative pattern, then the next test contrast would be increased (restorative pattern made easier to see).

During each trial, an auditory tone indicated stimulus onset, and auditory feedback was provided to indicate whether the response was correct or incorrect. The image to which the target was added was randomly selected at the beginning

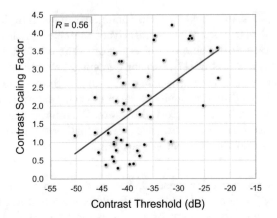

Fig. 3. Scatterplot of CT vs. CS. In general, harder-to-detect patterns (high CTs) require greater contrast scaling factors (high CSs).

of each trial, and observers were instructed to examine all three images before responding. Response time was limited to within 7 s of stimulus onset, during which all three images remained visible. After 30 trials, the contrast threshold was estimated as the 75%-correct point on a Weibull function, which was fitted to the data from the trials. RMS contrast (in dB) was used as the contrast metric.

3 Results

3.1 Contrast Thresholds vs. Contrast Scaling Factors

The contrast thresholds (CTs) obtained for each stimulus were combined across all subjects to obtain raw mean CTs and raw standard deviations of those means. These preliminary values were then used to discard outliers: Any subject's CT value which was outside of one raw standard deviation of the respective raw mean was considered an outlier. The remaining thresholds were then averaged, resulting in outlier-corrected mean CTs and standard deviations.

Figure 3 shows a scatterplot of average CTs vs. optimal contrast scaling factors (CSs) from [11]. As can be seen from the plot, there is a weak relationship between CT and CS. The relationship indicates that when the CT is low, the CS also tends to be low, and vice-versa, suggesting that a more easily detectable pattern requires lesser contrast to effect the maximal quality improvement as deemed by the subjects. The correlation coefficient of the linear regression is $R = 0.56$; there is a significant effect of CT on CS: $t(48) = 4.68, p < 0.001$. This finding supports the hypothesis that the visibility of a restorative pattern is one factor which determines it optimal contrast scaling factor.

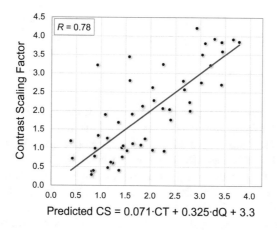

Fig. 4. Scatterplot of predicted CS (via the linear combination of CT and dQ) vs. actual CS.

3.2   Influence of Quality Improvement Potential on the CT-vs.-CS Relationship

Another factor which can possibly influence the optimal contrast scaling factor is the degree to which the restorative pattern visually matches the original texture. For many of the poorly matched restorative patterns, subjects generally preferred a very small contrast scaling factor. Thus, in some cases, the CS could also be low, not necessarily because the CT is low, but simply because of the poor visual match (i.e., the noise could not be shaped properly via the employed texture-synthesis method).

Although we did not measure subjective ratings of visual match, we did measure in [11] the quality improvement scores (dQs). These scores represent how much of an improvement the restorative patterns afforded when added to the distorted textures (at the optimal contrast factor) as judged by the subjects. The range of these dQ scores was 0–10, with zero denoting no improvement, and 10 denoting as high-quality as the original. By using both CT and dQ as regressors, we repeated the linear regression to predict CS. The regression revealed a significant effect of CT: $t(47) = 4.09, p < 0.001$; and a significant effect of dQ: $t(47) = 5.97, p < 0.001$. (When the interaction term CT $\times$ dQ was tested, only dQ was significant at the 0.05 level, suggesting that CT and CT$\times$dQ account for much of the same variance.) The resulting scatterplot for this regression is shown in Fig. 4. The correlation coefficient of this linear regression is $R = 0.78$.

Overall, these results support the hypothesis that the visibility of the restorative pattern can help determine the optimal contrast scaling factor. More specifically, when the CT is low, the CS tends to be low; however, when the CT is high, the CS may be high or low, depending on the dQ. If the dQ is high, then a high CT tends to correspond to a high CS. If the dQ is low, then the CS tends to be low, regardless of the CT. Thus, in order for a future algorithm to properly predict the optimal CS factor, it would be necessary to not only predict the CT, but also predict either the dQ or perhaps some measure of the degree of visual match between the restorative pattern and the original texture.

4   Conclusions

The use of perceptually shaped noise patterns for enhancement of compressed textures is a promising approach, but one which requires the patterns be set to proper contrasts, values which are texture-/image-specific, and thus need to be automatically predicted. In an effort to shed light on what factors might be useful for such a prediction, in this paper, we have presented the results of a psychophysical experiment designed to investigate the relations between the visual detectability of shaped-noise patterns, and the optimal subjective contrast scaling factors for the patterns when used to improve compressed textures. We hypothesized that patterns which have higher contrast detection thresholds (CTs) also require higher contrast scaling factors (CSs).

Using 50 textures collected from 15 images, we measured CTs in the current study and compared them with CSs measured in our previous study. The results of our experiment supported the hypothesis that low CT implies low CS, and vice-versa, but only for those patterns which could markedly improve the quality. This finding suggests that a future algorithm designed to predict the optimal CS for a given texture should take into account both the quality-improvement potential of the pattern (e.g., the degree to which it visually matches the texture) and the visibility (CT) of that pattern when it is applied to the texture.

References

1. Chandler, D.M., Lim, K.H.S., Hemami, S.S.: Effects of spatial correlations and global precedence on the visual fidelity of distorted images. In: Rogowitz, B.E., Pappas, T.N., Daly, S. (eds.) Proceedings of the SPIE Human Vision and Electronic Imaging XI, San Jose, CA (2006)
2. Golestaneh, S.A., Chandler, D.M.: Algorithm for JPEG artifact reduction via local edge regeneration. J. Electron. Imaging **23**(1), 013018 (2014). https://doi.org/10.1117/1.JEI.23.1.013018
3. Kayargadde, V., Martens, J.: Perceptual characterization of images degraded by blur and noise: experiments. J. Opt. Soc. Am. A **13**(6), 1166–1177 (1996)
4. Kshibuchi, Y., Aoki, N., Inui, M., Kobayashi, H.: Improvement of description in digital print by adding noise. J. Soc. Photogr. Sci. Technol. Jpn. **66**(5), 471–480 (2003)
5. Kurihara, T., Aoki, N., Kobayashi, H.: Analysis of sharpness increase by image noise. In: Proceedings of SPIE, vol. 7240, pp. 724014–724014-9 (2009). http://dx.doi.org/10.1117/12.806078
6. Olmos, A., Kingdom, F.A.A.: McGill calibrated colour image database. http://tabby.vision.mcgill.ca
7. Portilla, J., Simoncelli, E.P.: A parametric texture model based on joint statistics of complex wavelet coefficients. Int. J. Comput. Vis. **40**(1), 49–70 (2000). https://doi.org/10.1023/A:1026553619983

8. Shen, M.Y., Kuo, C.C.: Review of postprocessing techniques for compression artifact removal. J. Vis. Commun. Image Represent. **9**(1), 2–14 (1998). https://doi.
   org/10.1006/jvci.1997.0378
9. Wan, X., Kobayashi, H., Aoki, N.: Improvement in perception of image sharpness
   through the addition of noise and its relationship with memory texture. In: Proceedings of SPIE, vol. 9394, pp. 93941B–93941B-11 (2015). http://dx.doi.org/10.
   1117/12.2082922
10. Watson, A.B., Pelli, D.G.: Quest: a bayesian adaptive psychometric method. Per-
    cept. Psychophys. **33**, 113–120 (1983)
11. Yaacob, Y.M., Zhang, Y., Chandler, D.M.: On the perceptual factors underlying
    the quality of post-compression enhancement of textures. In: Proceedings of Human
    Vision and Electronic Imaging 2017 (2017)

Full Dynamics and Optimization of a Controllable Minimally Invasive Robot for a Soft Tissue Surgery and Servicing Artificial Organs

Grzegorz Ilewicz[✉] and Andrzej Harlecki

Faculty of Mechanical Engineering and Computer Science,
University of Bielsko-Biala, Bielsko-Biala, Poland
gilewicz@ath.bielsko.pl

Abstract. Minimally invasive robots are currently used during the operations of human body in the entire world. There are various important mechanical quantities in the designing process of a medical robot structure, such as the dynamic safety factor, which is reckoned in this following article. Matlab program was used to calculate the torque in every joint with a DC motor, controlled with PID regulator for a given trajectory. Trajectory allows the effector to move in the tunnel of a tissue inside the patient's chest (tunnel was obtained by using 3D Slicer computer program and CT scan diagnostics). Subsequently, the FEM (finite element method) was applied to enumerate transient conditions during the deformation of robots' structure (RRRS). Numerical experiment of multi-objective optimization is introduced in the work, where two criteria, during the calculation, are very significant: first natural frequency and dynamic safety factor for multibody effector of a medical robot in motion, taking into account the inertia, damping, and stiffness reactions. The Pareto optimum for these criteria is calculated by using a genetic algorithm. The mini robot effector is finished with a scalpel.

Keywords: Medical robot · Dynamical safety factor · Transient state
DC motor · PID controller · Genetic algorithm MOGA · CT scan

1 Introduction

Medical robots are used during the operations of human bodies or for servicing artificial organs and have structures with serial chain or closed kinematical chain. An example of a robot with a closed kinematical chain is the American da Vinci robot, which is most often used in minimally invasive operations. In the following article, the structure of a medical robot with an open kinematical chain and RRRS configuration is discussed, enabling minimally invasive surgeries, similar as the da Vinci robot. However, it has a bigger advantage of being able to reach the back wall of the operated organ and has wider possibilities in servicing the artificial organs. This opens a way to new surgical procedures, which due to its kinematic structure cannot be performed by a robot with a closed structure.

© Springer Nature Switzerland AG 2019
G. Laukaitis (Ed.): INTER-ACADEMIA 2018, LNNS 53, pp. 204–211, 2019.
https://doi.org/10.1007/978-3-319-99834-3_27

This benefit of the RRRS robot results from the configuration, i.e. the open kinematic chain. The effector of a mentioned robot can be equipped with various operational instruments (scalpel, needle, tongs) or it can be a tool for repairing artificial organs, e.g. the artificial heart. A calculation and solid model of a robot for servicing artificial organs that allows to reach the back wall of the operated organ is shown in Fig. 1.

Fig. 1. A robot for servicing an artificial organ with an RRRS configuration

The safety of using a medical robot is the most important criterion adopted during its construction. Next, the aim is to obtain an adequate stiffness of the kinematic chain and to minimize the criterion, which is the dynamical factor of safety of the robot. All these mechanical criteria must be met by assuming the appropriate functionality of a medical robot imposed by the surgeon's requirements. In the early design phase, surgical procedures should be determined by a team of surgeons and the functionality of the structure should be adapted to their implementation.

One of the most vital requirements set for medical robots is the value of the safety factor, which is referred in this article as the ratio of a yield point to equivalent stress according to the von Mises-Huber's hypothesis.

The main purpose of the article is to create a model that allows to determine the dynamical safety factor for a structure that is in motion, taking into consideration the transient states by using the finite element method. Such a created model is subjected to the process of searching for the optimal dimensions of the pipe walls from which the robot's parts are created. The Kriging method of fitting the meta-model to the data from numerical experiments and the genetic algorithm MOGA are used for this reason. Then the optimal strength structure is imported into the Matlab/Simulink environment, where are examined its dynamics in the system with servomotor models of subsequent degrees of freedom and control systems based on PID controllers.

Strength parameters of the elements of a medical robot, in the statics, were appointed in using the finite element method in the following works [1, 2]. The endurance model that allows to receive the optimal criterion, which is the stiffness while obtaining the minimum weight of the robot, is shown in work [3]. The mathematical equation of an eigenvalue problem are also shown in this article. The model of a medical robot with a control system is presented in work [4], while the model with drive and control systems is created in the article [5], where results were gained in the form of drive moments obtained in the joints of the structure.

The usage of a program to create calculation models based on photos from a computed tomography in medical robotics is shown in work [6]. In the model systems of the movement of medical robots, frictional interactions in their joints are also essential. This kind of influence on the dynamics of the analyzed robot was not taken into consideration in this particular work. However, these issues were under consideration during the analysis of other systems with a structure built on the basis of open kinematic chains, i.e. the ones that robots contain, for example [7].

The medical robot model created in this work makes it possible to obtain courses of mechanical quantities close to reality, which is why it is a useful tool that allows in a no experimental way to receive information about the robot's states during its work in the operating field.

Simulation in a numerical way of the operating room is carried out by using a chest model created on the basis of 2D data from the computed tomography diagnostics as shown in Fig. 2.

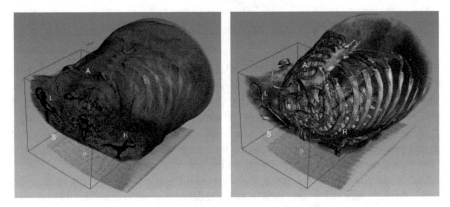

Fig. 2. Model of patient's chest from CT scans

2   Numerical FEM Optimization Model

A discrete model of medical robot was generated by using a mesh finite element method used in structural strength researches. A mesh was formed by discretizing the geometric model with tetrahedral Solid186 elements. The numerical model used in calculations has 5906 elements. Finite element mesh is presented in Fig. 3.

Fig. 3. Mesh of finite elements of a medical robot model

The robot's movement took place in the first two degrees of freedom with velocities equals 1 [rad/s]. A force 20 [N] was applied to the effector in the opposite direction to the scalpel blade.

In order to acquire the optimal strength of the structure, the target function was determined where the criteria are the first natural frequency and the dynamic safety factor:

$$f(\{\boldsymbol{d}\}) = \{f_1\{\boldsymbol{d}\}, f_2\{\boldsymbol{d}\}\} \tag{1}$$

$$4 \leq d_1 \leq 9.9 \ [\text{mm}]$$

$$4 \leq d_2 \leq 9.9 \ [\text{mm}]$$

$$60 \leq f_1\{\boldsymbol{d}\} \leq 400 \ [\text{Hz}]$$

$$4 \leq f_2\{\boldsymbol{d}\} \leq 5$$

knowing that:

$f_1\{\boldsymbol{d}\}$ - is a vector of the first natural frequency,
$f_2\{\boldsymbol{d}\}$ - is a vector of the dynamical factor of safety,
d_1, d_2 - dimensions, from which result the wall thickness of the first and second links.

The solution of optimization model was made by using the MOGA optimization algorithm.

Figures 4 and 5 present the response surfaces gained by Kriging method for the data from numerical strength experiments. The acquired graphs show the meta-model, which shortens the calculation time of the MOGA optimization algorithm.

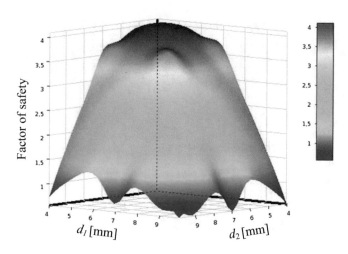

Fig. 4. Response surface for dynamical safety factor depending on the sought geometry dimensions

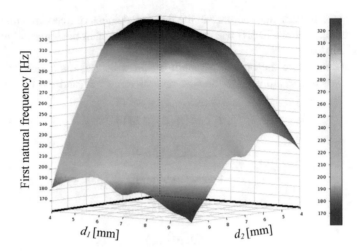

Fig. 5. Response surface for first natural frequency depending on the sought geometry dimensions

The following values of searched wall dimensions were obtained using MOGA multi-objective genetic algorithm, i.e. $d_1 = 4.68$ [mm] $d_2 = 5.06$ [mm] with optimal criterion of dynamic safety factor equal 4.08 and optimal criterion of first natural frequency equal 329.83 [Hz].

3   Model of DC Motor and Control System

The analytical equations of the DC motor model in continuous time have the following form:

$$\frac{d}{dt} \cdot \begin{bmatrix} \frac{d\omega}{dt} \\ i \end{bmatrix} = \begin{bmatrix} -\frac{b}{J} & \frac{k}{J} \\ -\frac{k}{L} & -\frac{R}{L} \end{bmatrix} \cdot \begin{bmatrix} \omega \\ i \end{bmatrix} + \begin{bmatrix} 0 \\ \frac{1}{L} \end{bmatrix} \cdot U \tag{2}$$

$$y = \begin{bmatrix} 1 & 0 \end{bmatrix} \cdot \begin{bmatrix} \omega \\ i \end{bmatrix} \tag{3}$$

knowing that:

   k - back electromotorical force coefficient,
   i - current,
   L - inductance,
   ω - rotational velocity,
   R - resistance,
   b - viscous coefficient,
   U - voltage.

The model (2, 3) has been attached to each connector of the mechanical model of a medical robot with an open kinematical chain. In Figs. 6, 7, and 8 block diagrams of the mechanical system, DC motor, and PID controller of the medical robot model can be scrutinized.

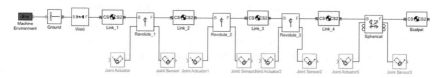

Fig. 6. Mechanical model of a medical robot

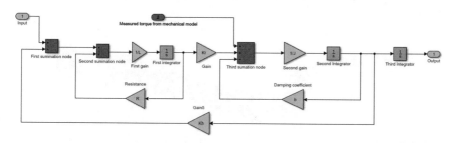

Fig. 7. Block model of a DC motor

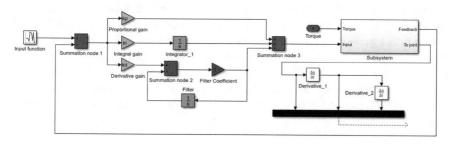

Fig. 8. Block model of PID controller and DC motor

The addition of drive and control systems to the mechanical model makes it possible to state that the considered dynamics of the system will be close to the dynamics of the real object.

This type of an approach is methodically proper, where first the optimal numerical results are obtained and then on the basis of such data a prototype of the device is formed.

The presented systems enable smooth control of the movement of the robot's geometric model with given mass distributions.

4   Results

Figure 9 illustrates the distribution of the equivalent stress in transient states for achieved optimum. The calculation of this value is possible after solving the equation of dynamics specifying the inertial and stiffness, and the damping in the robot's internal structure. The influence of deformations from previous iterations on the occurring subsequent deformations of the robot structure is also taken into account. Stress distributions are shown in Fig. 9, using the stress isolines. The largest stress concentrations are observed in the robot's joints, where the angular displacement values have been applied, which were realized by the model during the simulation test. These are the areas that are the most exposed to significant displacements under the influences of forces and loads appearing in motion, and the most influencing on deterioration of the two basic quality parameters of the medical robot's work, i.e. the accuracy of positioning and repeatability.

Fig. 9. Stress reduced according to von Mises-Huber's hypothesis in transient states

Figure 10 illustrates Pareto fronts for the dimensions, first natural frequency and dynamical safety factor obtained during MOGA numerical experiment. The applied MOGA algorithm is defined as better when looking for global optimizers than for example, the gradient - based algorithms NLPQL and MISQP, which are ideal for searching the local optima. The Pareto fronts for first natural frequency and dynamical safety factor, shown in Fig. 10, are the fronts of non-dominated solutions obtained by the iterative evolutionary method. In Fig. 10, the best set of numerical data during MOGA experiment is marked in blue, while the worst set of numerical data is marked in red.

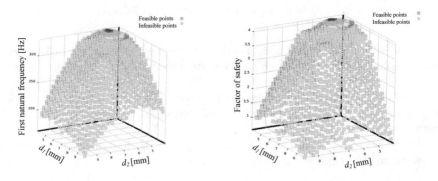

Fig. 10. Pareto fronts for dimensions, first natural frequency and dynamical factor of safety

For the mechanically optimal structure, drive moments were specified in subsequent joints by using a mechanical model designed in the Matlab/Simulink environment similar as in the work [5].

The obtained results allow concluding about the dynamic states in joints and places of a medical robot, where dangerous values of safety coefficient occur during its movement.

5 Conclusions

The article presents a full dynamics model of a medical robot, which made it possible to determine the optimal value of the dynamic safety factor, bearing in mind the inertia effects. The created model opens a door to find the answer about safety, which is a superior validity, the use of a medical robot in the operating field for given inputs. Receiving the answer to this question is crucial for a medical robot due to the necessities of state standards that decide about the possibility of using a medical robot for operations on human bodies. A DC motor model was added for the created optimal endurance model, enabling the robot's structure to be driven in subsequent joints and control system based on the PID controller, which resulted in obtaining a mechatronic model faithfully simulating the behavior of the real object. Dynamic characteristics of driving moments for a given extortion were also achieved. In sequent scientific works, a LuGre tribological model of resistance for motion will be added to the dynamics model, presented in this paper.

References

1. Yuan, H., Li, Z., Wang, H., Song C.: Static modeling and analysis of continuum surgical robots. In: IEEE International Conference on Robotics and Biomimetics (ROBIO) (2016)
2. Elsayed, Y., Vincensi, A., Lekakou, C., Geng, T., Saaj, C., Ranzani, T., Cianxchetti, M., Menciassi, A.: Finite element analysis and design optimization of a pneumatically actuating silicone module for robotic surgery applications soft robotics, vol. 2 (2014)
3. Ilewicz, G.: Natural frequencies and multi-objective optimization of the model of medical robot with serial kinematical chain. Adv. Intell. Syst. Comput. **519**, 371–379 (2017)
4. Ilewicz, G.: Multibody model of dynamics and optimization of medical robot to soft tissue surgery. Adv. Intell. Syst. Comput. **393**, 129–134 (2016)
5. Ilewicz, G.: Modeling and controlling medical robot for soft tissue surgery and servicing the artificial organs. In: Proceedings of the 17th Mechatronika 2016, pp. 102–107. IEEE, Prague (2016)
6. Ki, S., Tan, Y., Deguet, A., Kazanzides, P.: Real-time image-guided telerobotic system integrating 3D slicer and the da vinci research kit. First IEEE Int. Conf. Robot. Comput. (2017). https://doi.org/10.1109/IRC.2017.73
7. Harlecki, A., Urbaś, A.: Modelling friction in the dynamics analysis of selected one-dof spatial linkage mechanism. Meccanica **52**, 403–420 (2017)

Aerial Image Registration for Grasping Road Conditions

Kyoji Ogasawara[✉] and Hitoshi Saji

Graduate School of Integrated Science and Technology, Shizuoka University,
3-5-1, Johoku, Naka-ku, Hamamatsu, Shizuoka 432-8011, Japan
ogasawara.kyoji.17@shizuoka.ac.jp

Abstract. It is useful to obtain road traffic information quickly for disaster rescue and remediation activities. Utilizing aerial images for determining road traffic conditions is an effective means that is not affected by the situation on the ground. In particular, helicopters can be used to not only grasp road traffic conditions but also aid rescue activities. Moreover, systems that transmit real-time images to ground stations, such as helicopter relay systems and helicopter satellite systems, have attracted attention. In fact, such systems widely contributed to the rescue activities after the Kumamoto earthquakes that occurred in April 2016. However, there is a large difference between the fixation point and position of photographing because aerial images are obtained obliquely with respect to the ground surface, which remains a problem. In this study, we identify the fixation point without utilizing any large special equipment by only using information from the global positioning system (GPS) and a map. Ultimately, we aim to use the results of registration between the images and maps to grasp road traffic conditions.

Keywords: Aerial images · Image geo-registration · Map

1 Introduction

1.1 Background

Large-area information and road traffic information around regions affected by disaster are used to secure routes for rescue vehicles, urgent material transport vehicles, and disaster victims. Aerial images are an effective means of collecting traffic information because they can convey disaster information over large areas and are not affected by the situation on the ground. An increasing number of municipalities and national organizations have adopted helicopter TV relay systems (called heli-tele in Japanese) or helicopter satellite communication systems (called heli-sat in Japanese) for collecting information from aerial images during large-scale disasters. These systems can transmit real-time images obtained from a helicopter to stations on the ground. Such images are used to grasp the extent of damage and aid rescue activities. Helicopters have higher mobility than satellites and also contribute to rescue activities immediately after the occurrence of disasters. However, it is difficult to specify accurate fixation point with only information from images. To address this problem, a system of registering the

© Springer Nature Switzerland AG 2019
G. Laukaitis (Ed.): INTER-ACADEMIA 2018, LNNS 53, pp. 212–218, 2019.
https://doi.org/10.1007/978-3-319-99834-3_28

frames of a moving image to a map has been proposed and used to determine fixation point [1]. However, registration is difficult in the case of bird's-eye-view images, in which the ground surface is photographed obliquely and the fixation point of the photograph is greatly different from the photographing position. Therefore, a technique for accurately registering bird's-eye-view images to a map is required. Here, image registration means that two or more different images are superimposed on the basis of a relation that exists between them.

1.2   Purpose of This Study

It is necessary to estimate the fixation point to accurately register bird's-eye-view images to a map, which requires high-accuracy sensors. However, helicopters should not be equipped with large special equipment since more load should be allotted to rescue activities. Therefore, accurate registration should be realized using the minimum necessary amount of information. Iigura et al. [2] achieved versatile registration using information from only images and GPS. However, the estimated fixation point included a certain amount of error, and some time was required to correct this error. Therefore, improved estimation accuracy of the fixation point or the correction of estimation results is required to apply their registration method to real-time use. In addition, the method of improving estimation accuracy of the fixation point and the area coverage was realized by Kojima et al. [3], who used the quantitative road features. In this method, however, a visual check is required by an operator to ensure accurate registration because the estimated fixation point includes error. This makes the method more difficult to be used by people from other regions or dispatched from the government to help rescue activities in large-scale disasters.

In this study, we propose an automated registration method that can also be used by people unfamiliar with the local area. The position and orientation of an airplane are estimated by the methods of Iigura et al. and then information on buildings and roads from a map is used to find a key frame from moving images that is suitable for registration. Finally, using the key frame as a base image, registration and correction are performed on the basis of quantitative road features by the methods of Kojima et al.'s method.

2   Outline of Research

2.1   Problem Setting

The conditions of the airplane flight and image projection are summarized in Table 1.

Table 1. Conditions of flight and photography

Condition	Notes
Altitude of airplane 150 m or higher during photography	According to the Civil Aeronautics Act, an airplane must fly 150 m or higher above the ground surface or buildings on the ground
Ground surface to be photographed can be approximated by a plane	Differences in topographical height should be small. No mountains or areas with high concentration of tall buildings
Ground surface is photographed unidirectionally	No arbitrary points photographed multidirectional. Each point is photographed only once
GPS data can be obtained	GPS data can be obtained for each time-series frame

2.2   System Configuration

The proposed system consists of three modules. The first is a module for estimating the position of the airplane and the orientation of the camera, which are required for the registration of bird's-eye-view images. The second is a module for extracting the key frame suitable for registration on the basis of the map information and the fixation point obtained from the position/orientation estimation module. The third is a module for registering the images to a map using the quantitative road features and the key frame. The position/orientation estimation module is based on the estimation of the airplane motion by homography decomposition in addition to Iigura et al.'s method. The image–map registration module is based on Kojima et al.'s method.

Position/Orientation Estimation Module. The pseudo-orthogonal space used by Iigura et al. is constructed by an optical flow assuming that the motion of an airplane is linear for a certain period of time. Therefore, nonlinear motion causes an error in the estimation of the fixation point. In homography decomposition, the motion existing between time-series images is decomposed into rotational and translational components. The rotational component is fed back as an error to reduce the error in the estimation of the fixation point. The internal parameters of the camera are required for homography decomposition and are calculated using the pseudo-orthogonal space.

Key Frame Extraction Module. The focal length estimated using the pseudo-orthogonal space includes error, resulting in error in the estimation of the fixation point. To reduce this error, a frame is matched to a rasterized map. In this study, we focus on road edges with the aim of performing registration at the level of the ground surface. However, the results of correction using edges greatly depend on the condition of each frame and the map of the surrounding area. To solve this problem, a key frame that is easily corrected is extracted using a map of the surrounding area. The map information is used to calculate the road width, the number of buildings, and the entropy of buildings (hereafter, building entropy). The frame with the most reliable result of edge correction among the moving images is selected and used as the key frame.

Image–Map Registration Module. In the key frame extraction module, the key frame is extracted and results are corrected using edges. In this study, we focus on regions of road to grasp road traffic conditions and perform registration of the images having regions of road. Mosaicked images based on the key frame are created by the method of Lin et al. [4] and registered using quantitative road features. A Gaussian mixture model (GMM) is used to extract the regions of road.

3   Proposed Method and Procedures

3.1   Estimation of Fixation Point in Pseudo-orthogonal Space

Bird's-eye-view images have perspective distortion and should be transformed to orthogonal images for accurate registration. To construct an orthogonal space required to transform bird's-eye-view images to orthogonal images, vanishing points and the focal length must be known. The pseudo-orthogonal space proposed by Iigura et al. is constructed using the traveling direction of the airplane and the normal direction of the ground surface in the normal camera space, and can be constructed even though the focal length is unknown. In this study, the position of the airplane and the orientation of the camera are estimated on the basis of the pseudo-orthogonal space, employing a system that estimates the true state from the observations including errors. This system adopts a Kalman filter to correct the data on the basis of GPS and estimate the focal length.

3.2   Extraction of Key Frame

The result of edge correction and the information from a map of the surrounding area are used to extract a key frame. Because perspective distortion is mostly removed from the images by the position/orientation estimation module, we aim to minimize the parallel translation errors between the images and the map. Therefore, template matching using edges of roads and buildings extracted from the images and the map is performed. The roofs of buildings are a major source of noise during matching between the images and the map. Moreover, the sides of buildings are projected in images taken obliquely with respect to the ground surface, greatly affecting the result of registration. We use the number of buildings and building entropy to estimate the regions that are unsuitable for registration, for example, the regions with concentrated buildings and large buildings such as factories. Furthermore, wide roads are detected because the existence of wide roads is expected to increase the registration accuracy.

Estimation of Density of Buildings. Kumagai et al. [5] examined a method of analyzing the density of buildings using satellite image data. They used entropy from thermodynamics and statistical mechanics as an index of the layout of buildings when quantitatively analyzing their density. The entropy (E) of buildings is given by

$$E = \frac{1}{N}\log\left\{\frac{N!}{N_1! \times N_2! \times \cdots \times N_k!}\right\} \tag{1}$$

$$= \sum_{j=1}^{k} f_j \log f_j (j = 1, 2, \cdots k) \tag{2}$$

Here, j is the identification number of a pixel group recognized as a building, N_j is the number of pixels constituting the building, $N = N_1 + N_2 + \cdots + N_k$, and $f_j = N_j/N$. The entropy is higher in areas where more buildings are concentrated, whereas it is lower in areas with fewer buildings. Moreover, the entropy tends to decrease in areas having large buildings with many pixels. In this study, the entropy is used along with the number of buildings to extract a frame covering an area where buildings are not concentrated and few large buildings exist.

4 Registration Based on Quantitative Road Features

Extraction of Roads Using GMM. Road regions are extracted from bird's-eye-view images. This extraction is performed for a block of mosaicked images of the subsequent frame. To deal with the difference in pixel values of different roads and misjudgments caused by noise, the pixels are classified according to the GMM. The GMM is the result of linear integration of some Gaussian distribution, as below, and stochastically classifies pixels by applying the probability model to the data distribution. The maximum likelihood estimation of the mean, variance, and mixing coefficient of the Gaussian distribution is performed using the expectation–maximization (EM) algorithm. In this study, a model was made to learn the pixel values of road and nonroad regions in advance.

Registration by Gauss–Newton Method. The road regions extracted from the image block are registered to the road regions on the map. Kojima et al. minimized the residual error of registration by the Gauss–Newton method. In this method, the Hessian matrix is approximated by the square of the Jacobian matrix to reduce the calculation load. Kojima et al. also performed a coarse-to-fine search using an image pyramid to solve the convergence to local solutions and further reduce the calculation load. This method is effective because the calculation load can be reduced to some extent, increasing the processing speed. Hence, we adopted this method in this study.

5 Experiments

5.1 Experimental Data

In the experiments, we used aerial images taken from a helicopter of Hamamatsu Fire Department, Hamakaze, and a map provided by the Geospatial Information Authority of Japan. GPS data were included in the aerial images and partly extracted for use.

5.2   Results of Registration

To evaluate the proposed method, we performed experiments on registration between the aerial images and the map at three locations. The map information was used after rasterizing the vector data in shapefiles. Figure 1 show the results of registration obtained by the proposed method.

The registration errors are evaluated for each intersection because the proposed method is assumed to be used to grasp road traffic conditions. The error for each intersection is calculated and the mean and maximum errors is determined. A summary of the results of each location is shown in Table 2.

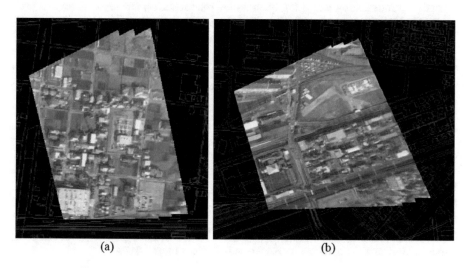

(a)   (b)

Fig. 1. Result of registration: (a) Experiment 1, (b) Experiment 2

Table 2. Evaluation results

Experiment	Mean of error [pixel]	Maximum error [pixel]
1	3.0	9
2	17.5	59
3	24.6	48

6   Discussion

- Error of registration
  Registration was performed using the quantitative road features of mainly wide roads. The results showed that automated registration was achieved without the visual confirmation of operators. As shown in Fig. 1(b), there were few registration errors for wide roads, although large errors occurred for the area where roads were shaded by buildings.

- Improved extraction of key frame
  The road width, the number of buildings, and building entropy on the map of the surrounding area were used to extract the key frame. However, we require quantitative features that will enable us to grasp the features of road conditions more accurately. For example, the layout of buildings, the degree of occupancy of buildings on a map, and the presence/absence of curved roads can be examined as quantitative features to increase the accuracy of registration.

- Expansion of target area
  To grasp road traffic conditions during large-scale disasters, registration should be performed over wider areas. In this study, we performed the registration of wide areas by mosaicking and using quantitative road features. However, there is a limit to the detection accuracy of roads from bird's-eye-view images that include shadows of buildings. Hence, we will examine the use of simultaneous localization and mapping (SLAM) as a means of obtaining information on the camera and the ground surface from images. We will also examine the use of a digital surface model (DSM) to apply our method to areas with differences in height, such as areas of mountain forests and tall buildings.

7   Conclusion

In this study, we constructed an image-map registration system using only image and GPS data. This system can perform automated registration without manual operation by operators and is suitable for people unfamiliar with the local area. However, more accurate registration over wider areas is required to grasp road traffic conditions during large-scale disasters. In future work, we will address these issues.

Acknowledgements. This work was supported by JSPS KAKENHI Grant Numbers JP18K04657.

References

1. Nonoyama, Y., Ijiri, M., Maeda, K.: A system that displays the position of a helicopter images. Mitsubishi Denki giho **78**(5), 367–371 (2004)
2. Iigura, K., Saji, H.: Oblique aerial image registration from a movie and GPS Data. J. Remote Sens. Soc. Jpn. **35**(3), 155–172 (2015)
3. Kojima, R., Saji, H.: Geo-registration of Consecutive Aerial Images to Grasp Road Situation. ITS World Congress, Montreal (2016)
4. Lin, Y., Yu, Q., Medioni, G.: Map-enhanced UAV image sequence registration. In: IEEE Workshop on Applications of Computer Vision (WACV 2007), p. 15 (2007)
5. Kumagai, K.: Development of an analysis method on the building density assuming the utilization of high-resolution satellite data. Data **54**(700), 111–121 (2002)

Synthetic Light Interference Image Analysis Algorithm in Wedge Interferometer

Jakub Mruk[(⊠)]

Faculty of Mechatronics, Warsaw University of Technology, A. Boboli 8,
Warsaw, Poland
mruk.kuba@gmail.com

Abstract. This work presents a project of new interference image analysis software. The main idea was to create a new algorithm that uses the position of functions local maxima to detect the wedges position change. The purpose of this software is to determine the possibility of using flat-parallel plate interference to determine the position of the wedge plate in relation to the laser beam, with the use of CCD camera.

Keywords: Algorithm · Interference · Image analysis · Wedge

1 Introduction

Interference in a wedge plate is a well-known phenomenon that is used commonly in measuring applications [1]. One of its metrological uses is examination of shape deviations of reference elements, e.g. surface of gauge blocks, optical flats, or micrometer anvil [2]. Interference in a glass window is used for collimation of laser beams.

This paper examines the ability to measure the rotation of a small-wedge plate, using intensity distribution analysis of two interfering beams of different wavelengths [3]. Wedge plate is used to obtain interference fringes in space beyond axis of the beam, what allows to direct it on the CCD camera.

Intensity of two interfering beams reflected of front and inner surface of fixed thickness wedge plate depends on the difference in optical path. This path depends on wedge geometry and the angle at which the beams fall. With the use of wedge with very small angle (around 5″) and small angle rotation (+2°) it is possible to direct the beam onto the photodetector and acquire compaction of interference fringes, what allows detection (interference fringes have period around 0.1–1 mm).

2 Theory

Rotation of optical element causes change in spatial distribution of intensity. In addition, applying light of two different, dominant wavelengths reveals the overlap of different wave interfering frequencies (Fig. 1).

© Springer Nature Switzerland AG 2019
G. Laukaitis (Ed.): INTER-ACADEMIA 2018, LNNS 53, pp. 219–224, 2019.
https://doi.org/10.1007/978-3-319-99834-3_29

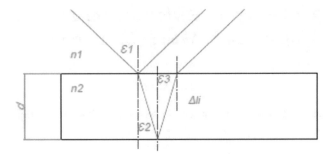

Fig. 1. The path of beam in optic element (ε_1 – angle of incidence, ε_2 –angle of refraction, ε_3 – angle of reflection, d – wedge thickness, n_1 – refractive index of air, n_2 – refractive index of glass, Δl_i – optical path)

For the case in question, difference in optical path Δli, in glass, of same source beams, can be simplified and given in form:

$$\Delta l_i = 2d/\cos\left(\arcsin\left(\frac{n}{n_{Li}} \cdot \sin(\varepsilon_1)\right)\right) \tag{1}$$

ε_1 – angle of incidence, n – refractive index of air, n_L – refractive index of glass, d – wedge thickness.

When rotating the wedge in relation to the light beam the angle of incidence is also the angle of rotation.

Intensity at each point of cross-section can be written as a composition of two wavelength beams:

$$I(\Delta l) = I_0 + \sum_{i-1}^{n=2} A_i \cdot \cos((2 \cdot \pi \cdot \Delta l_i)/\lambda_i) \tag{2}$$

I_0 – constant intensity, λ_i – wavelength, A – constant responsible for the intensity of i – number of light source, Δl_i – optical path difference.

Intensity at point of interference image depending on the rotation angle (angle of incidence), based on Eqs. (2) and (1) gives modulation with variable frequency beat period (Fig. 2).

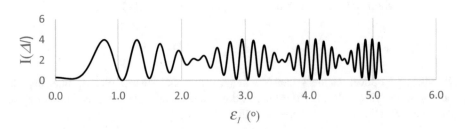

Fig. 2. Intensity versus angle of incidence

Due to discrepancy of beams and angle of the wedge in a plane of photodetector, the fringes are observed on the surface. Intensity distribution is modulated by Gaussian function of beam intensity. In a constant position of the wedge, intensity distribution is approximately symmetrical to the Gaussian envelope, and the distance between beats is constant (Fig. 3). Difference from dispersion introduces some asymmetry. This factor, however, is not taken into account at this stage of research.

position (pixel)

Fig. 3. Intensity distribution of interfering beam (without dispersion)

Rotation of the wedge causes a change in beat density, seen in Fig. 3, according to Eq. (2), presented in Fig. 2.

An attempt was made to measure the change in the distance of two adjacent local maxima caused by rotation of the wedge. The measurement was carried out in the program, based on the analysis of fringes distribution image. This measurement will be used to determine the wedge rotation based on changes in the image.

3   Experimental Setup

Optical setup consisting of four main elements: source of light (2 laser diodes LD1 and LD2 of different wavelengths), beamsplitter to join the beams BS, wedge interferometer W (wedge plate mounted on angular table), Camera CCD operating as photodetector. Interference of beams reflected of front and inner surfaces of the wedge was observed.

Interferometer setup generated interference fringes composition of two different wavelengths. That allowed to detect local maxima and minima of the periodic function, which modulated the interference fringes [3] (Fig. 4).

The setup used one wedge plate, 7 mm thick, with 0.05 rad wedge angle. Two diodes, with wavelengths of 660 nm and 780 nm powered by external, stabilized power supply. CCD camera, connected to laptop, recorded video in 1920 × 1080 pixels.

a)

b)

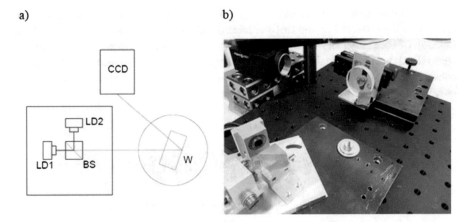

Fig. 4. (a) Measuring setup with synthetic light source, wedge and detector and (b) technical implementation of the system

3.1　Algorithm for Analysis of the Interference Distribution

An all-new algorithm analyzes interference image based on density distribution of interference fringes. Software was developed in LabView environment.

Algorithm determines position change of local extremum. Maximum of the function acts as reference point – absolute position of wedge in relation to the beam. The change of its position can be calculated into wedge rotation. Figure 5 shows front panel of the software, with two-beam interference image in the left, intensity distribution in top right side and intensity histogram in the bottom left side, which is used to adjust the setup.

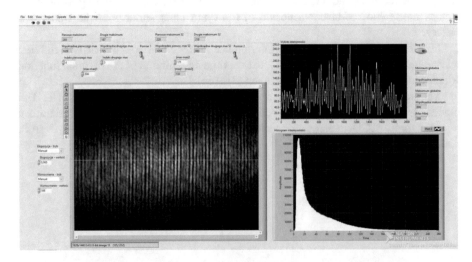

Fig. 5. Front panel – used for the analysis of interference distribution

The algorithm can be divided in three main parts. First of them is responsible for the connection between CCD camera and computer. After the connection is established it helps to configure the photodetector. Then a line is drawn, along which intensity of every pixel is saved, and plotted into two graphs. First one being the histogram of intensity (right bottom corner of front panel), second one being intensity diagram (right top corner of front panel).

Second part of the algorithm divides captured image into number of smaller intervals. The program finds maximum of intensity in every of those intervals, and then groups (clusters) the value of the intensity with pixel number on which it occurred and the interval number. After that the cluster is sorted, the distance between two biggest intensities is calculated and saved (in pixels).

After rotating the wedge plate the software detects the maximum values of intensity in the same two intervals the previous two global maximums appeared. Then it calculated the distance between them. In the last step the distances of second and third step are subtracted, to define if the distance between the maximum increased or decreased.

Knowing the size of a pixel and the difference in pixel distance, we can convert distance in pixels into millimeters, and then into angular deviation.

4   Results and Discussions

Initial tests have shown that the programs algorithm operates properly. The image of modulated intensity distribution was achieved (Fig. 5). Rotation of the wedge caused displacement of the fringes and change in their density. A change in the maxima distances could be detected (Fig. 6).

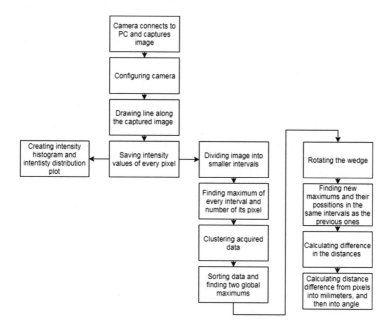

Fig. 6. Block diagram of algorithm

Occasional excessive errors were observed, which were equal to a multiple of the wavelength. Diode laser beam axis overlapping and their proper collimation were difficult to obtain, although all necessary regulations were provided. The right distribution was acquired for fringes density corresponding to incidence angle equal to 0.174 rad. Interference fringes in the observation field was 60.

5   Conclusions

The source of errors was the spatial distribution of synthetic light. The resolution of the camera was insufficient for correct mapping. Number of pixels per fringe - 23 – was also too small to estimate the right maximum.

In further works it is planned to improve adjustment of the setup and select sources with superior beam intensity distribution parameters.

References

1. Bong Song, J., Lee, Y.W., Lee, I.W., Le, Y.-H.: Simple phase-shifting method in a wedge-plate lateral-shearing interferometer. Appl. Opt. **43**(20), 3989–3992 (2004)
2. Yoshizawa, T.: Handbook of Optical Metrology: Principles and Applications, pp. 426–428. CRC Press, London (2003)
3. Iwasińska-Kowalska, O., Dobosz, M.: A new method of non-contact gauge block calibration using a fringe-counting technique: II. Experimental verification. Opt. Laser Technol. **42**, 149–155 (2010)

A Flood-Fill-Based Technique for Boundary Closure of Kidney Contours in CT Images

Tomasz Les[1]([⊠]), Tomasz Markiewicz[1,2], Mirosław Dziekiewicz[2], and Małgorzata Lorent[2]

[1] Warsaw University of Technology, Warsaw, Poland
lest@ee.pw.edu.pl
[2] Military Institute of Medicine, Warsaw, Poland

Abstract. The article presents an innovative method of kidney recognition in computed tomography (CT) images. Kidney cancer is one of the most common causes of death. Over 300,000 people die per year from this disease. A fast and correct diagnosis of neoplastic lesions in computed tomography images allows to choose the proper method of treatment. This article presents innovative and unique methods of kidney recognition in CT images. The proposed methods are based on morphological operations, shape analysis, geometrical coefficients calculations as well as the directional operation of flood fill with automatic selection of the stop criterion. The article presents also an innovative method of closing the boundary of an unrecognized kidney. Application of fast and effective algorithms for an automatic kidney shape recognition allows to make a 3D reconstruction of the kidney model. The use of algorithms to improve visualization of CT scans allows more accurate diagnosis by specialists. The system for supporting kidney cancer diagnosis presented in the article has been tested to assess the quality of kidney shape recognition. The recognition results of the shape of the kidney by the automatic system are comparable to the results obtained by a human expert and the accuracy of the diagnosis is at the level of 86%. Despite the difficult task, it was possible to obtain satisfactory results of the kidney shape recognition.

Keywords: Flood fill · Shape analysis · Geometrical coefficients
Brightness correction

1 Introduction

The article presents an innovative method of recognizing the shape and location of kidney. Kidney cancer is currently one of the deadliest types of cancer. The treatment still gives unsatisfactory results. The detection of kidney cancer is at low level. Often, kidney cancer is recognized during a routine medical examinations of other organs. Due to the severity of the disease, the invasive removal of the kidney is necessary. Therefore a fast and effective detection of neoplastic lesions in computed tomography images is crucial to increase the possibility of effectively curing the cancer. Current techniques of computer image analysis allow to automatically find the location of a kidney and to determine its exact contour. Defining the exact shape of a kidney allows

© Springer Nature Switzerland AG 2019
G. Laukaitis (Ed.): INTER-ACADEMIA 2018, LNNS 53, pp. 225–232, 2019.
https://doi.org/10.1007/978-3-319-99834-3_30

specialists to assess the degree of neoplastic changes in the kidney. Most of the known algorithms for automatic image processing can correctly determine the position of the kidney and calculate its contour at the same time [1–3]. Unfortunately, a correct determination of boundary characteristic still causes a problem. This is due to the fact that kidney often overlaps other organs, e.g. liver. This makes the boundary between the organs often blurred. Therefore typical solutions based on the analysis of pixel brightness in the gray-scale will fail. This article presents an innovative method of identifying the location of kidney along with a technique of closing the kidney's boundary. The proposed solution is based on the flood fill algorithm with modified stop criterion depending on distance and brightness function. Despite the difficult task of identifying the kidney contour, it was possible to obtain satisfactory results in the system tests.

2   Problem Statement

The problem of supporting automatic medical diagnostics of a kidney cancer is reduced to the analysis of tomographic images. Computed tomography is a common medical examination of internal organs. Tomography in the kidney cancer study is a scan of a certain abdominal segment. As a result of the study, a set of several dozen to several hundred digital grey-scale images is obtained. Each image represents individual fragment of internal organs. Analyzing manually such a large amount of data is a tedious and a time-consuming task and involves a high risk of mistake. Computer analysis allows to increase the accuracy of the medical examination and to reduce the diagnostic time. There is a need to develop methods for rapid image analysis so that the examination result is available immediately. In this work, the middle abdominal section is chosen, with clearly visible kidney. Unfortunately, in this fragment the boundary of kidney is often blurred or overlapped with other organs. Therefore, the identification of the kidney is quite a simple task, but it is difficult to find precisely the characteristic of the kidney boundary. Most known algorithms skip the step of closing the boundary. In this work, the technique of closing the boundary of kidney is presented. An exemplary CT image in which the problem of overlapping organs is clearly visible is shown in Fig. 1.

Fig. 1. An example CT image on which the problem of overlapped organs is visible

In Fig. 1, the left kidney (in the picture on the right side) is overlapped with liver. It's difficult to determine the precise boundary between the two organs. The algorithm developed and presented in this work allows to solve this problem. The proposed solution is universal and can also be used in other problems of automatic identification of objects in the image.

3 Techniques for Kidney Recognition

Determination of the kidney boundary in CT images is a complex process. In the first step, the image is scaled to a range of 0–255. The scaling uses the level of spinal cord's brightness (as the brightest point near kidney). In the next step, finding the edge with the Canny method is performed. Out of all the detected edges, the longest edge located closest to the points: A: $\{x: 402, y: 306\}$ and B: $\{x: 149, y: 294\}$ are selected.

Points A and B are selected empirically. They indicate the location where the left and right kidneys are most likely disposed. The typical algorithms of kidney detection generate incomplete boundary of kidney. Therefore, it is necessary to apply the procedure of closing the kidney boundary. This technique is presented in point 4.

The complete system for detecting areas occupied by kidney includes four main steps: (1) image scaling operations, (2) edge detection, (3) directional flood fill and, finally, closure of the kidney boundary.

4 Closing the Boundary

The edge detection algorithm, based on Canny method, successfully recognizes most of the kidney contours. A frequent problem, difficult to avoid, is a discontinuous kidney boundary. This is due to the high variability of brightness intensity of pixels in the kidney area. The level of pixel brightness values between two opposite points on the kidney edge may vary by up to 30%. This problem is visible in Fig. 2.

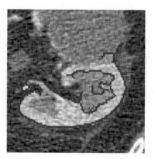

Fig. 2. Characteristic interruptions in the kidney boundary

In order to eliminate this problem, an algorithm of closing the kidney edge based on the flood fill was prepared. In this work, two algorithms were used - standard flood fill - used to find end points of the edge - and extended algorithm - directional flood fill - used to close the boundary of kidney.

4.1 Flood Fill

Flood fill is a popular algorithm used in graphic programs to fill closed areas of a bitmap with a specified color. This algorithm uses three parameters: the initial position, the color to be changed and the new color. An important parameter affecting the result of algorithm is a number of directions of propagation. The most commonly used parameters are 4 (up, down, left and right) and 8 (up, down, left, right, top-left corner, top-right corner, bottom-left corner and bottom right-hand corner) [4].

4.2 Directional Flood Fill

Directional flood fill is an innovative extension of the standard flood fill algorithm. The presented algorithm was developed to solve the problem of discontinuity character of the detected kidney boundary. The main idea of the presented algorithm, as well as the standard flood fill, is to fill a coherent area with a selected color. The aim of the presented algorithm is to fill the subset of points lying in the path between selected points A and B with a specified color. A visualization of the directional flood fill algorithm is presented on Fig. 3.

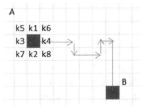

Fig. 3. Visualization of the directional flood fill algorithm

The found path does not have to be the shortest path between points $A = \{x_a, y_a\}$ and $B = \{x_b, y_b\}$ according to meaning of the Euclidean distance. There are two functions defined: distance - $d(x)$ and pixel brightness - $b(x)$. Both parameters are taken into account when filling the next point.

From point A there is a possibility to expand in 8 directions, marked as points: $K_1 \ldots K_8$. Forereach point: $K_i = \{x_i, y_i\}, i \in \{1 \ldots 8\}$ distance function:

$$d(K_i) = \sqrt{(x_i - x_b)^2 + (y_i - y_b)^2} \tag{1}$$

and brightness function:

$$b(K_i) = |v(x_i, y_i) - v(x_p, y_p)| \tag{2}$$

are defined. In the above formula $v(x, y)$ means brightness value of image in point: $\{x, y\}$. Point: $P_{j-1} = \{x_p, y_p\}$ is the point visited in the previous algorithm step, where the current step is marked as j.

Finally, for each point: K_i the minimum of following function is calculated:

$$M = \min(2 * d(K_i) + b(K_i)) \qquad (3)$$

Finally, the point P_{j+1} for which the minimum M is met is selected.

4.3   The Kidney Boundary Closing Algorithm

This section presents a complete algorithm for closing the boundary of kidney. This algorithm uses the technique of traditional flood fill and directional flood fill described in Sect. 4.2. In the first step, all end points of edges of the kidney boundary are found with the use of the flood fill algorithm. From the central point (found by calculating the center of figure) of the kidney boundary, 8 directional growth is made into all possible directions. Points that cannot be further expanded are marked as endpoints. In Fig. 4. an example fragment of the kidney boundary with the selected end points A and B has been presented.

Fig. 4. Fragment of the kidney boundary with visible selected endpoints (the most distant from each other)

In the next step, two end points that are as far away as possible from each other are found. From these two points, directional flood fill is made, taking into account the criterion of previously defined distance and brightness functions. After n steps of algorithm (in test n = 20) we check if:

$$dist(A_n, B) < dist(B, A) \qquad (4)$$

where A_n are points in n step of algorithm. If the result of Eq. (4) is true, then the algorithm is continued. When the result of (4) is false, then next starting point is selected and the procedure is repeated. The Eq. (4) allows to check if the growth is carried out in the right direction (towards the end point – B). Figure 5 presents the result of the flood fill algorithm going in the wrong direction – (b) and the result of the flood fill going in the correct direction (c).

The developed technique allows to close the boundary of the kidney, taking into account the probable further direction of the kidney edge.

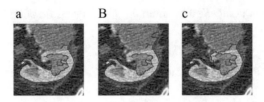

Fig. 5. An example of a flood fill algorithm that goes in the wrong direction (b) and the correct direction (c)

5   Tests and Results

In order to assess the quality of recognition of the kidney boundary by the computer system, an automatic tests were performed. The shape of the kidney was compared to the shape marked manually by an expert. 48 different CT images were analysed in the tests. In each image there are two kidneys. Altogether, 96 kidney boundaries were analysed. In order to assess the automatic recognition of the kidney, binary masks of kidneys were prepared - on the basis of the boundary determined by the automatic system and the human expert. The masks have been cut on all sides and imposed on each other. Finally, the number of pixels for which both masks take the same value (1 or 0) has been counted. System accuracy (*acc*) is defined by the following formula:

$$acc = TP + TN/(TP + FP + TN + FN) \qquad (5)$$

where *TP* (true positive) and *TN* (true negative) indicates the number of pixels that match the compatibility of both masks, while *FP* indicates the number of pixels not belonging to the expert mask but detected by system and *FN* indicates the number of pixels belonging to expert mask but not detected by system [5]. All 48 analysed images were grouped into 10 groups containing from 8 to 11 kidney masks. In addition, two groups has been defined: separable organs and overlapped organs.

Additionally, the average BDE error (The Boundary Displacement Error expressed as the average displacement error of boundary pixels between two segmented images) was calculated based on work [6].

The numerical results of the experiment are presented in Table 1.

Analysing Table 1, we can see that it can be seen that the main error of both groups is similar and is respectively 86.11% for separable organs and 84.97 for overlapped organs. The result of a complete kidney edge detection system is shown in Fig. 6.

6   Conclusion

The process of manual detection of the kidney contour in the image is a time-consuming and laborious task. The result of the analysis may vary depending on many external factors such as lighting, the place where the test was performed or the psychophysical condition of the person carrying out the analysis. Detecting boundaries is a manual task requiring a lot of practice. The computer performs this task in a much

Table 1. Numeric results of the experiment

Group	Case Separable organs	Case Overlapped organs	Group	Case Separable organs	Case Overlapped organs
1	90,67	93,39	6	90,67	83,70
2	87,87	89,47	7	87,87	84,85
3	88,78	84,67	8	88,78	80,76
4	80,56	85,76	9	80,56	82,45
5	82,67	83,02	10	82,67	81,65

measure	Average values Separable organs	Average values Overlapped organs
acc	86.11	84.97
BDE	13.76	12.87

Fig. 6. The result of a complete kidney edge detection system

shorter time, always with the same accuracy, regardless of external factors. This is very important because kidney cancer diagnosis requires the analysis of multiple scans and each scan should be analysed with the same accuracy. The automatic system proposed in the work allows to find a contour with a satisfactory accuracy of 86%. The developed technique of closing the kidney edge can be used also in other works and other systems, increasing their accuracy.

Acknowledgment. This work has been supported by the National Science Centre (2016/23/B/ST6/00621 grant), Poland.

References

1. Canny, J.: A computational approach to edge detection. IEEE Trans. Pattern Anal. Mach. Intell. **8**, 679–698 (1986)
2. Song, H., Kang, W., Zhang, Q., Wang, S.: Kidney segmentation in CT sequences using SKFCM and improved GrowCut algorithm. In: IEEE International Conference on Bioinformatics and Biomedicine (BIBM 2014) Belfast, UK, 2–5 November 2014 (2014)

3. Tsagaan, B., Shimizu, A., Kobatake, H., Miyakawa, K., Hanzawa, Y.: Segmentation of kidney by using a deformable model. In: International Conference on Image Processing, Thessaloniki, Greece, vol. 3, pp. 1059–1062 (2001)
4. Torbert, S.: Applied Computer Science, 2nd edn, p. 158. Springer, Berlin (2016). https://doi.org/10.1007/978-3-319-30866-1
5. Song, H., Kang, W., Zhang, Q., Wang, S.: Kidney segmentation in CT sequences using SKFCM and improved GrowCut algorithm. In: IEEE International Conference on Bioinformatics and Biomedicine (BIBM 2014): Systems Biology
6. Allen, Y., John, W., Yi, M., Shankar, S.: Unsupervised segmentation of natural images via lossy data compression. J. Comput. Vis. Image Underst. 110(2), 212–225 (2008)

Soft Computing Techniques and Modeling, Multimedia and e-Learning

A Hybrid Neuro-Fuzzy Algorithm
for Prediction of Reference Evapotranspiration

Amir Mosavi[1,2,3](✉) ⓘ and Mohammad Edalatifar[4]

[1] Institute of Automation, Kando Kalman Faculty of Electrical Engineering,
Obuda University, Budapest 1431, Hungary
amir.mosavi@kvk.uni-obuda.hu
[2] Department of Computer Science, Norwegian University of Science and
Technology, Trondheim, Norway
amir.mosavi@ntnu.no
[3] Institute of Advanced Studies Koszeg, iASK, Koszeg, Hungary
[4] Department of Electrical Engineering, Arak Branch, Islamic Azad University,
Arak, Iran

Abstract. In this study, a hybrid algorithm of adaptive neuro fuzzy inference system (ANFIS), particle swarm optimization (PSO) and principle component analysis (PCA) is utilized to predict the reference evapotranspiration (ET0). The accuracy of the computational model is evaluated using four statistical tests including Pearson correlation coefficient (r), mean square error (MSE), root mean-square error (RMSE), and coefficient of determination (R2). The results show that the ET0 can be estimated with an acceptable accuracy trough combination of PCA and ANFIS. Moreover, the result indicated that the ANFIS model can be simplified via reducing dimensionality of the input data.

Keywords: Neuro-Fuzzy · Reference evapotranspiration (ET0)
Principle component analysis · Particle swarm optimization · Prediction
Forecasting · ANFIS

1 Introduction

Irrigation and water resources management need accurate estimation and prediction of reference evapotranspiration (ET0) [1]. The FAO Penman–Monteith equation [2] is adopted and applied worldwide as a reference equation for estimation ET0 [3]. Nowadays, soft computing methods have been introduced and applied in estimation and forecasting ET0. A vast number of Artificial Neural Network (ANN) methods of intelligent computational, have been utilized as a tool for modelling nonlinear processes such as ET0 [4]. Gene expression programming (GEP) [5] has been applied in symbolic regression, time series prediction and in the field of hydrological modeling as well as ET0 prediction. Support vector machine (SVM) [6] is widely used as a machine learning algorithm to predict ET0 [7]. Here, we adaptive neuro-fuzzy inference system (ANFIS) which is a kind of ANN. Earlier, Keskin et al. [8] also used ANFIS to predict the evaporation in a similar application considering the climatic data. Furthermore, Karimaldini et al. [9] applied ANFIS for the daily modeling of ET0 considering the arid conditions. The literature, includes novel machine learning methods to find the most

© Springer Nature Switzerland AG 2019
G. Laukaitis (Ed.): INTER-ACADEMIA 2018, LNNS 53, pp. 235–243, 2019.
https://doi.org/10.1007/978-3-319-99834-3_31

influential ET0 parameters [10–15]. They conclude that duration of daily sunshine is the most influential parameter. Yet, the actual vapor pressure minimum, along with the air temperature are considered as the further parameters for ET0 estimation. The parameters of ANFIS are membership functions that they must be adjust. PSO, a stochastic optimization technique, is used to adjust ANFIS parameters [16, 17]. The PSO determines parameters of optimization problem in least iteration [18]. The data set which is used in this study have 8 features. When the data set have numerous features, the network parameters and their training phase become complex. In this case, the PCA as an effective statistical procedure, can be applied to extract features and reduce dimensions of data set. Further, we demonstrate that the combination of PCA and ANFIS is a promising approach. Moreover, it will be shown that the data which is reduced dimension with PCA can be used instead of data with original dimensions without any significant change in estimation results.

2   Data Set and Methodology

2.1   Data Set

The data set is gathered from two cities of Iran, at two meteorological stations of Tehran and Zanjan. It includes the meteorological data, sampled monthly for the period of 1981–2011, including actual vapor pressure (e_a), sunshine hours (n), wind speed, and minimum (T_{min}) and maximum (T_{max}) air temperature. Table 1 includes the geographic information of the cities. The mean annual T_{max} & T_{min}, respectively, varies from 13.22 to 16.19 °C and from 4.9 to 9.1 °C. Furthermore, the mean annual n varies for the candidate locations from 153.6 to 182.5 h. In addition, the range of mean annual e_a is from 0.91 to 1.5 kPa.

Table 1. Information of the meteorological cities of the data set

City	Elevation (m)	Latitude (N)	Longitude (E)
Tehran	1250	35°42′	51°24′
Zanjan	1659	46°40′	48°29′

2.2   Principal Component Analysis

The Principal Component Analysis (PCA) [19] is a fundamental statistical for extracting features and reducing the dimensions of data set. It moves data that they may have correlation to an uncorrelated space called feature space. In fact, PCA transforms related variables to set of uncorrelated variables that are called PCs (Principle Components) [20, 21]. Because data set with large sample data need a massive space for storage and need a complicated processing, it is important to reduce data dimensions. To reduce dimension, we can omit some PCs that they have small variance, because PCs are uncorrelated.

Assume $X_{m\times n}$ is a matrix of data to calculate PCA, m is a number of dimensions and n is a number of sample data. First data are normalized ($\bar{X}_{m\times n}$), and then covariance matrix of $\bar{X}_{m\times n}$ is calculated as:

$$R_{m \times m} = \frac{1}{n-1} \bar{X} \bar{X}^T \tag{1}$$

Eigenvectors and eigenvalues of covariance matrix (R) can be calculated using singular value decomposition (SVD). Because eigenvalue is a rectangular diagonal matrix, diagonal matrix is saved in column vector shown with U. In next step, U is sorted in descending shown with λ then eigenvector is sorted according λ rows. Furthermore, $\bar{X}$ can be projected in feature space:

$$Y_{m \times n} = V_{m \times m} \times X_{m \times n} \tag{2}$$

where V is eigenvector of R and Y is projection of $\bar{X}$ in feature space. Y has size equal to $\bar{X}$.

Reducing Dimensionality with PCA

It is possible to reduce size of Y with elimination some PCs but it must be noted that some of information are lost. The value of λ elements has close relationship with information of Y. If row 'a' of λ has big value, the row 'a' of Y has big information. Then a row, PC, of Y can be deleted if corresponding column of λ has small value. There are a number of criteria to calculation remain information after elimination some PCs. Cumulative Percent Variance (CPV) is used in this study [22, 23]:

$$CPV\% = \frac{\sum_{i=1}^{k} \lambda(i)}{\sum_{i=1}^{n} \lambda(i)} * 100 \tag{3}$$

where k is a number of PCs that remained. For example, if CPV is 98.5% and k is 5, it means with remain 5 first row of λ and corresponding eigenvector's row, 98.5% of information are remained and 1.5% of information are lost. After delete some rows of eigenvector ($V_{reduced}$), reduced data ($Y_{reduced}$) is calculated with Eq. 4.

$$Y_{reduced} = V_{reduced} \times X \tag{4}$$

Further, ANFIS is used to build the network. The parameters are membership functions (MF) to be determined. The adjustment of Sugeno fuzzy inference system (FIS) parameters which is a difficult and timely task is conducted through PSO.

2.3 Particle Swarm Optimization (PSO)

PSO has a number of particles that each particle is a data with two features, position and value [12, 24]. Position is input data (X) and value is an answer of fitness function (Y). PSO tries to move particles to better position in a number of iterations. Always, better position and value of particles are kept in Xpbest matrix and Ypbest vector, respectively. In iteration for a particle, if value of particle becomes better, new values are replaced in Xpbest and Ypbest. In addition, the better values of past iterations are always kept in Xgbest and Ygbest. In other words, Xgbest is the best position found by PSO until current iteration. New position of a particle is influenced by four parameters:

current position, current velocity, distance between current position and Xpbest as well as distance between current position and Xgbest is calculated by:

$$X_i^{t+1} = w_i X_i^t + cp \times rand() \times (Xpbest_i - X_i) + cg \times rand() \times (Xgbest_i - X_i) \quad (5)$$

Where ith particle is shown with index i. Index t shows iteration number. W_i is an inertia factor, cp is an individual-best acceleration factor and cg is a global-best acceleration factor. Values of W_i, cp and cg are changed linearly during iterations. Also, *rand ()* is a function that produce a number between zero to one randomly. Answer of PSO is Xgbest and Ygbest in last iteration. An algorithm for PSO is shown in algorithm (1).

Algorithm (1): algorithm for PSO

Input: number of particles (npart), number of iterations (niter), initial and final values of the individual-best acceleration factor (cbi, cbf), initial and final values of the global-best acceleration factor (gbi,gbf), initial and final values of the inertia factor (wi,wf) and fitness function.

Output: position of particle with better fitness value.

Start algorithm:

Select npart particles over input space uniformly at random and save in matrix X.

Calculate fitness value of all particles and input in Y vector.

Calculate increase value of the inertia factor (w), the global-best acceleration factor (gb) and of the individual-best acceleration factor (cb) in each iteration with below equations:

$$\Delta cg = \frac{gbf - gbi}{niter}, \ \Delta cb = \frac{cbf - cbi}{niter}, \ \Delta w = \frac{wf - wi}{niter}$$

Initialize w, cb, cg, Xbest and Ybest with below equations:

$$w = wi, \ cb = cbi, \ cg = cgi, \ Xbest = X, \ Ybest = Y$$

For niter iterations **do, For** all particles **do**

Update position of particle using

$$X_i^{t+1} = w_i X_i^t + cp \times rand() \times (Xpbest_i - X_i) + cg \times rand() \times (Xgbest_i - X_i)$$

Calculate value of fitness function for particle in new position ($Y_i = fitness(X_i)$).

If $Y_i < Ybest_i$ **then**

$Ybest_i = Y_i$

$Xbest_i = X_i$

End

End for

The position and value of a particle that it has minimum value in Ybest save in Xgbest and Ygbest, respectively.

Increase w, cb and cg by $\Delta w, \Delta cb \ and \ \Delta cg$, respectively.

End for

End algorithm

When parameters of network are adjusted, it is possible that overtraining happens. It means the network can answer to train data, yet with lower quality. To prevention of overtraining, portion of data are used in train process called validation data. Validation data are used in train process, but they aren't used for training. If output of fitness function for input validation data during several iterations doesn't improve or become worse, Xpbest and Ypbest for this iteration must not be change. Therefore, the main part of PSO algorithm is changed according algorithm (2).

Combination of PSO and ANFIS

The purpose of this study is to determine the exact parameters of ANFIS. It is possible that determined value of ANFIS parameters with PSO. In order to use PSO, a fitness function is needed. Fitness function is a function that gets input data (X) i.e. position in

PSO algorithm and return error value (Y). Therefore, a function that its inputs are parameters of ANFIS and input data as well as its outputs are MSE of output data and output of fuzzy network is used in this study as a fitness function.

Algorithm (2): Main part of PSO algorithm that it is modified for prevention of overtraining

invalidCounter = 0, For niter iterations do, For all particles do
Update position of particle using equation:
 	
$$X_i^{t+1} = w_i X_i^t + cp \times rand() \times (Xpbest_i - X_i) + cg \times rand() \times (Xgbest_i - X_i)$$
 	
Calculate value of fitness function for particle in new position ($Y_i = fitness(X_i)$).
If $Y_i < Ybest_i$ then
    $Ybest_i = Y_i$
    $Xbest_i = X_i$
End
        If output of fitness function for input validation data is worse do
            invalidCounter = invalidCounter + 1
        Else
            invalidCounter = 0
            XbestValid = Xbest
      YbestValid = Ybest
        End if
        If InvalidCounter = maximum invalidation do
The position and value of a particle that it has minimum value in YbestValid save in Xgbest and Ygbest, respectively.
    Go to end algorithm.
End if
      End for
The position and value of a particle that it has minimum value in Ybest save in Xgbest and Ygbest, respectively.
    Increase w, cb and cg by $\Delta w, \Delta cb \ and \ \Delta cg$, respectively.
  End for

Notice that input and output data are constant and each time that call fitness function they are given to function. It means particle of PSO are FIS parameters. This fitness function flowchart is shown in Fig. 1.

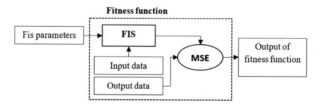

Fig. 1. Fitness function used in this study

2.4   FAO–56 Penman–Monteith Equation

Allen et al. [2] proposed the FAO–56 Penman–Monteith (PM) equation to estimate reference evapotranspiration. The PM equation is presented as follow:

$$ET_0 = \frac{0.408\Delta(R_n - G) + \gamma \frac{900}{T + 273} U_2(e_s - e_a)}{\Delta + \gamma(1 + 0.34U_2)} \tag{6}$$

where ET0 is equal to the reference evapotranspiration (mm day-1), and G = soil heat flux density (MJ m-2 day-1). The Δ = slope of the saturation vapor pressure function (kPa oC-1). Furthermore, the γ = psychometric constant (k Pa oC-1), Rn = net radiation (MJ m-2 day-1), T = mean air temperature (oC), and U2 = average daily wind speed. In this study the ET0 is calculated using the application provided by Gocic and Trajkovic [25].

3   Results

In this study, the monthly values of vapor pressure (e_a), maximum (T_{max}) and minimum (T_{min}) air temperature, minimum (RH_{min}) and maximum (RH_{max}) relative humidity, sunshine hours (n) and wind speed (U_2) during the period 1980–2010 were used for generating PSO and FIS models. Data sets is divided into training, testing and validation. The training data set includes observed data measured for the period 1982–1995, the testing data set is for the period 1996–2005, while the validation data set is for the period 2006–2015. Tables 2, summarizes the parameters value i.e., standard deviation (σ) and mean (μ) for training, testing and validation data set.

3.1   Statistical Performance Evaluation

Here to evaluate the performance of PSO and ANFIS models, four popular statistical indices were utilized as follow:

$$RMSE = \sqrt{\frac{\sum_{i=1}^{n} (P_i - O_i)^2}{n}} \tag{7}$$

$$MSE = \sqrt{\sum_{i=1}^{n} (P_i - O_i)^2} \tag{8}$$

$$R^2 = \frac{\left[\sum_{i=1}^{n} (O_i - \bar{O}_i) \cdot (P_i - \bar{P}_i)\right]^2}{\sum_{i=1}^{n} (O_i - \bar{O}_i) \cdot \sum_{i=1}^{n} (P_i - \bar{P}_i)} \tag{9}$$

$$r = \frac{n\left(\sum_{i=1}^{n} O_i \cdot P_i\right) - \left(\sum_{i=1}^{n} O_i\right) \cdot \left(\sum_{i=1}^{n} P_i\right)}{\sqrt{\left(n\sum_{i=1}^{n} O_i^2 - \left(\sum_{i=1}^{n} O_i\right)^2\right) \cdot \left(n\sum_{i=1}^{n} P_i^2 - \left(\sum_{i=1}^{n} P_i\right)^2\right)}} \tag{10}$$

where Oi = predicted ET0 values obtained by PSO and FIS models, Pi = FAO-56 PM values of ET0, and n = the number of training and testing data.

Table 2. Parameters of training, testing and validation of data set (1982–2015)

	Tmin (oC)		Tmax (oC)		RHmin (%)		RHmax (%)		ea(kPa)		U2(ms-1)		n(h)	
Training														
Station name	μ	σ	μ	σ	μ	σ	μ	σ	μ	σ	μ	σ	μ	σ
Tehran	8.00	0.60	16.81	1.01	55.30	2.87	79.90	3.15	1.03	0.05	1.66	0.22	167.09	11.81
Zanjan	6.07	0.70	16.03	1.28	59.56	4.14	84.67	2.54	1.04	0.07	1.98	0.26	167.55	12.68
Testing														
Station name	μ	σ	μ	σ	μ	σ	μ	σ	μ	σ	μ	σ	μ	σ
Tehran	8.52	0.71	17.29	1.13	56.00	3.91	77.58	2.83	1.07	0.05	1.61	0.24	171.51	13.65
Zanjan	6.55	0.66	16.82	1.22	60.01	4.00	84.77	2.45	1.10	0.04	1.86	0.32	180.97	14.26
Validation														
Station name	μ	σ	μ	σ	μ	σ	μ	σ	μ	σ	μ	σ	μ	σ
Tehran	9.50	0.50	18.13	0.65	54.88	2.60	76.15	2.47	1.10	0.04	1.81	0.11	182.49	13.14
Zanjan	7.25	0.30	17.54	0.85	60.20	3.59	85.15	1.81	1.15	0.03	1.73	0.11	189.03	21.62

3.2 Further Model Analysis

The setting of ANFIS and PSO that are used in this study are presented in [26]. Suppose every input of FIS has 10 Gaussian MF with 2 parameters each, every input would needs 20 parameters to adjust. This subject shows importance of reduced dimensionality. To avoid overtraining, 15% of data is used as validation data. The

Table 3. Results of performance analysis of network for RMSE, MSE, R2, and r

Station name	Dimension size	Training data				Testing data			
		MSE	RMSE	R^2	r	MSE	RMSE	R^2	r
Tehran	8	0.0727	0.2697	0.9696	0.9847	0.0905	0.3008	0.9636	0.9817
	7	0.0590	0.2429	0.9753	0.9876	0.0704	0.2653	0.9717	0.9858
	6	0.0706	0.2656	0.9706	0.9852	0.0511	0.2260	0.9794	0.9896
	5	0.0756	0.2750	0.9684	0.9841	0.0811	0.2848	0.9677	0.9837
	4	0.0705	0.2655	0.9705	0.9852	0.0940	0.3065	0.9621	0.9809
	3	0.1146	0.3385	0.9521	0.9758	0.0962	0.3102	0.9611	0.9804
	2	0.1172	0.3423	0.9510	0.9752	0.1798	0.4240	0.9292	0.9639
	1	0.1675	0.4092	0.9300	0.9644	0.1485	0.3854	0.9407	0.9699
Zanjan	8	0.0578	0.2404	0.9698	0.9848	0.0570	0.2387	0.9747	0.9873
	7	0.0562	0.2370	0.9720	0.9859	0.0906	0.3010	0.9618	0.9807
	6	0.0366	0.L913	0.9809	0.9904	0.0645	0.2540	0.9696	0.9847
	5	0.0469	0.2166	0.9754	0.9876	0.0639	0.2528	0.9692	0.9845
	4	0.0538	0.2320	0.9718	0.9858	0.0527	0.2295	0.9749	0.9874
	3	0.0558	0.2362	0.9708	0.9853	0.0618	0.2487	0.9707	0.9852
	2	0.1178	0.3432	0.9383	0.9687	0.1255	0.3543	0.9382	0.9686
	1	0.1954	0.4420	0.8989	0.9481	0.1787	0.4227	0.9146	0.9563

results of proposed method are shown in Table 3. For better comparison, ANFIS was trained with different dimensions of data. In other words, dimensions of data were reduced with PCA then parameters of ANFIS were optimized with PSO using output data of PCA. For checking the ability of the algorithm in response to new data, in addition to result of training data, result of testing data are presented in Table 3. Evaluated criteria have a few changes with reduce dimensions.

4 Conclusions

A hybrid algorithm of ANFIS, PSO and PCA is proposed to predict the ET0. Monthly data including minimum and maximum air temperatures, vapor pressure, minimum and maximum relative humidity, sunshine hours and wind speed have been used to develop the models. The data were given to the PCA to move them to an uncorrelated space and extract main features. The dimensions of the data were reduced with the PCA. Then the output data of the PCA were used to train the ANFIS. The results were shown that proposed approach has a good accuracy in term of r, R2, MSE and RMSE and it can be used to predict ET0. Results were also shown that many dimensions of the PCA outputs have not significant information. Therefore, through eliminating the dimensionality, a simpler ANFIS network with less parameters can be obtained.

Acknowledgment. Dr. Mosavi carried out this research during the tenure of an ERCIM Alain Bensoussan Fellowship Programme. Furthermore, the support and research infrastructure of Institute of Advanced Studies Koszeg, iASK, is acknowledged.

References

1. Pereira, L., et al.: Crop evapotranspiration estimation with FAO56: past and future. Agric. Water Manag. **147**, 4–20 (2015)
2. Allen, R.G., et al.: Crop Evapotranspiration. Guidelines for Computing Crop Water Requirements, vol. 56. FAO Irrigation and Drainage Paper, Roma (1998)
3. Raziei, T., Pereira, L.S.: Spatial variability analysis of reference evapotranspiration in Iran. Agric. Water Manag. **126**, 104–118 (2013)
4. Mattar, M.A.: Using gene expression programming in monthly reference evapotranspiration modeling: a case study in Egypt. Agric. Water Manag. **198**, 28–38 (2018)
5. Ferreira, C.: Gene expression programming: a new adaptive algorithm for solving problems. Complex Syst. **13**(2), 87–129 (2001)
6. Awange, J.L., et al.: Support Vector Machines (SVM). In: Mathematical Geosciences, pp. 279–320. Springer, Berlin (2018)
7. Gavili, S., et al.: Evaluation of several soft computing methods in monthly evapotranspiration modelling. Meteorol. Appl. **25**(1), 128–138 (2018)
8. Keskin, M., et al.: Estimating daily pan evaporation using adaptive neural-based fuzzy inference system. Theor. Appl. Climatol. **98**(1–2), 79–87 (2009)
9. Karimaldini, F., et al.: Daily evapotranspiration modeling from limited weather data using neuro-fuzzy computing technique. J. Irrig. Drain. Eng. **138**(1), 21–34 (2012)
10. Torabi, M., et al.: A hybrid machine learning approach for daily prediction of solar radiation. In: Advances in Intelligent Systems and Computing. Springer, Berlin (2018)

11. Torabi, M., et al.: A hybrid machine learning approach for daily prediction of solar radiation. In: Recent Advances in Technology Research and Education (2018)
12. Najafi, B., et al.: An intelligent artificial neural network-response surface methodology. Energies **11**(4), 860 (2018)
13. Moeini, I., et al.: Modeling the time-dependent characteristics of perovskite solar cells. Sol. Energy **170**, 969–973 (2018)
14. Mosavi, A., et al.: Reviewing the novel machine learning tools for materials design. In: Advances in Intelligent Systems and Computing, pp. 50–58 (2017)
15. Moeini, I., et al.: Modeling the detection efficiency in photodetectors with temperature-dependent mobility and carrier lifetime. Superlattices Microstruct. (2018)
16. Dineva, A., et al.: Fuzzy expert system for automatic wavelet shrinkage procedure selection for noise suppression. In: Intelligent Engineering Systems, pp. 163–168 (2014)
17. Mosavi, A.: The large scale system of multiple criteria decision making. Large Scale Complex Syst. Theory Appl. **9**(1), 354–359 (2010)
18. Mosavi A., et al.: Predicting the future using web knowledge: state of the art survey. In: Advances in Intelligent Systems and Computing, vol 660 (2018)
19. Jackson, J.E.: A User's Guide to Principle Components. Wiley, London (1991)
20. Mosavi, A., Varkonyi-Koczy, A.R.: Integration of machine learning and optimization for robot learning. In: Advances in Intelligent Systems and Computing (2017)
21. Mosavi, A., et al.: Industrial applications of big data: state of the art survey. Adv. Intell. Syst. Comput. **660**, 225–232 (2017)
22. Karimi, I., Salahshoor, K.: A new fault detection and diagnosis approach for a distillation column based on a combined PCA and ANFIS. In: Control and Decision, pp. 3408–3413 (2012)
23. Mosavi, A., et al.: Review on the usage of the multiobjective optimization package of modefrontier in the energy sector. In: Advances in Intelligent Systems and Computing, pp. 217–224 (2017)
24. Imani, M.H., et al.: Strategic behavior of retailers for risk reduction and profit increment via distributed generators and demand response programs. Energies **11**(6), 1–24 (2018)
25. Gocic, M., Trajkovic, S.: Software for estimating reference evapotranspiration using limited weather data. Comput. Electron. Agric. **71**(2), 158–162 (2010)
26. Kumar, M., et al.: Estimating evapotranspiration using artificial neural network. J. Irrig. Drain. Eng. **128**(4), 224–233 (2002)

The Numerical Study of Co-existence Effect of Thermal and Solutal Marangoni Convections in a Liquid Bridge

Chihao Jin[1], Atsushi Sekimoto[1], Yasunori Okano[1(✉)],
and Hisashi Minakuchi[2]

[1] Department of Materials Engineering Science, Osaka University, 1-3,
Machikaneyama, Toyonaka, Osaka 560-8531, Japan
okano@cheng.es.osaka-u.ac.jp
[2] Department of Mechanical Systems Engineering, University of the Ryukyus,
1 Senbaru, Nishihara, Okinawa 903-0213, Japan

Abstract. Marangoni convection is a flow along the interface between two fluids due to the variation of surface tension, which is mainly caused by temperature and/or concentration gradients, namely thermal and/or solutal Marangoni convection(s). During the crystal growth by the floating zone method, Marangoni convection may induce the striation and affect the quality of growing crystal. Therefore, it is necessary to understand the convective phenomenon and control it. In this study, a three-dimensional configuration of the half-zone liquid bridge in the crystal growth of Si_xGe_{1-x} was selected by establishing the temperature and concentration differences. The equations of continuity, momentum, energy and mass transfer were solved by the PISO algorithm in the OpenFOAM. Thermal and solutal Marangoni convections were set to flow in the opposite direction under zero gravity in the melt with different Marangoni numbers. Results have shown that when Ma_C is larger than $-Ma_T$, the flow is steady and axisymmetric, while Ma_C is smaller than $-Ma_T$, the flow is unsteady and irregular. As for the control of Marangoni convection, rotation of top and bottom plane in the liquid bridge was applied during the crystal growth. With different rotation speeds and directions, the suppression of Marangoni convection can be effectively realized by the appropriate forced rotation.

Keywords: Marangoni convection · Floating zone · Numerical simulation

1 Introduction

Semiconductor of Si_xGe_{1-x} is a new-developed material, which has many advantages, such as stable performance, fast data processing and lower energy consumption. Therefore, it is widely used in the manufacturing of optoelectronic devices in the field of information technology. However, when it comes to the crystal growth of such semiconductor material, Marangoni convection may occur along the interface in the melt zone due to the variation of surface tension mainly caused by temperature and/or concentration gradients, namely thermal and/or solutal Marangoni convection(s). In a floating-zone (FZ) system, between the feed and seed crystal, the liquid zone exists,

© Springer Nature Switzerland AG 2019
G. Laukaitis (Ed.): INTER-ACADEMIA 2018, LNNS 53, pp. 244–251, 2019.
https://doi.org/10.1007/978-3-319-99834-3_32

which is supported by internal surface tension and external electromagnetic force. By moving the heating coils, the single crystal grows from the seed crystal. In order to discuss the pure effect of co-existent thermal and solutal Marangoni convections, a floating half zone under zero gravity was considered as the liquid bridge in this study. As a non-crucible technique, there is less possibility to absorb oxygen into the growing crystal, which is free from the crucible contamination. Additionally, as the density of germanium is much larger than that of silicon, it is important to set the whole system under zero gravity to eliminate the natural convection led by the buoyancy and gravity since this research only sheds light on the mechanism of Marangoni convection. Although previous researches focused on thermal Marangoni convection only [1, 2], solutal Marangoni convection caused by concentration gradients was also considered in this study. As a kind of unsteady flow, Marangoni convection may induce the growth striation and lead to the concentration distribution, which means the uniformity of crystal will be affected. Apart from the internal mechanism, control of Marangoni convection should also be focused on. In this study, the forced convection of plane rotation was applied in the numerical simulation. During the Marangoni convection in the liquid bridge, plane rotations in the same and opposite directions were both considered. Marangoni convection can be changed or impaired by the plane rotations due to the centrifugal force, and to some degree, the external force can suppress the flow instability. In this study, co/counter-rotations were added to the simulation to observe their effects on Marangoni convection. Therefore, the objective of this research is to investigate the co-existence effect of thermal and solutal Marangoni convections, which flow in the opposite direction, and try to control the flow instability by plane rotation.

2 Numerical Method

During the crystal growth of Si_xGe_{1-x}, half of the liquid zone was considered as a liquid bridge as shown in Fig. 1. The liquid bridge has the free surface, and both top and bottom planes keep no-slip conditions. On the top plane, higher temperature was set and silicon was put. To the contrary, lower temperature and germanium was set on the bottom plane. In this case, the temperature and concentration gradients were in the opposite directions so that thermal and solutal Marangoni convections flow were opposite. Computational mesh of $Nr \times N\theta \times Nz = 30 \times 120 \times 40$ was used in the simulation. Continuity, momentum, heat and mass transfer equations were solved by the PISO algorithm by the OpenFOAM.

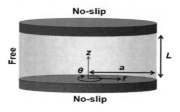

Fig. 1. Schematics of the half-zone liquid bridge [4]

Some simplifications are also made during the calculation as follows:

(a) The fluid is incompressible and Newtonian.
(b) The interfaces of the solid and liquid are flat.
(c) The whole system is under zero gravity to eliminate the natural convection.
(d) The liquid bridge remains cylindrical shape.

In order to describe the thermal and solutal Marangoni convections more directly, two non-dimensional numbers are introduced to the simulation, thermal and solutal Marangoni numbers (MaT and MaC), where $\frac{\partial \sigma}{\partial T}$ and $\frac{\partial \sigma}{\partial C}$ are surface tension coefficient of temperature and concentration, ΔT ($T_{bottom} - T_{top}$) and ΔC are temperature and concentration gradients, L is the length of liquid bridge, μ is viscosity and v is kinematic viscosity.

$$Ma_T = -\frac{\partial \sigma}{\partial T} \cdot \frac{\Delta TL}{\mu v} \tag{1}$$

$$Ma_C = \frac{\partial \sigma}{\partial C} \cdot \frac{\Delta CL}{\mu v} \tag{2}$$

Contrary to the previous study [3], thermal and solutal Marangoni convections are assumed to flow in the opposite directions by establishing the opposite-direction gradients. In this study, thermal Marangoni convection flows from the top to the bottom plane, however, solutal Marangoni convection flows from the bottom to the top plane. In this case, Ma_C was constant at 1072, and Ma_T varied from 0 to -3500. On the top plane, the temperature increased from 1300 K to 1310 K at the speed of 1 K/1000 s and kept constant at 1300 K on the bottom plane.

In terms of the control of the combined Marangoni convection, rotations of top and bottom planes were applied in the simulation. As a kind of forced convection, rotation can contribute to the suppression of Marangoni convection under appropriate speed and direction. Rotations of top and bottom planes were considered in the same and opposite direction with the same speed. The speed was set at 1 to 5 rpm. In this case, Ma_T was kept constant at 2857 and Ma_C at 1750, where thermal and solutal Marangoni convections were supposed to flow in the same direction.

Physical parameters used in the simulation are listed in Table 1.

Table 1. Parameters and properties of Si_xGe_{1-x} system

Property	Symbol	Value
Radius of the liquid bridge	a [m]	1.00×10^{-2}
Length of the liquid bridge	L [m]	5.00×10^{-3}
Aspect ratio	Asp	0.5
Thermal diffusivity	α [m²/s]	2.20×10^{-5}
Kinematic viscosity	v [m²/s]	1.40×10^{-7}
Diffusion coefficient	D [m²/s]	1.00×10^{-8}
Prandtl number	Pr	6.37×10^{-3}
Schmidt number	Sc	14.0
Rotation speed	ω [rpm]	1/2/3/4/5

3   Results and Discussion

3.1   Co-existence Effect of Thermal and Solutal Marangoni Convection

Figure 2 presents the temperature distribution with different thermal Marangoni numbers in the central r-z plane. Ma_T increased from 0 to -3500 and Ma_C was kept as 1072. When the temperature on the top plane increases at a steadily linear speed, the temperature distribution in the central r-z plane diffuses steadily from the top to the bottom plane. It reveals that the temperature distribution is steady and two-dimensional axisymmetric. The whole temperature field is almost steady.

Figure 3 shows the snapshots of concentration distribution with Ma_C = 1072 and increasing Ma_T in the r-θ plane at the height of 0.5L and central r-z plane of the liquid bridge. When there is no thermal Marangoni convection (Ma_T = 0), the flow pattern turns to be three-dimensional oscillating unsteady. The moving wave number shown in the concentration pattern of thermal and solutal Marangoni convections can be defined as azimuthal wave number, which can describe the flow instability. In the case (a), the azimuthal wave number equals four. When Ma_T increases from -350 to -1050, the concentration field becomes a two-dimensional axisymmetric steady flow. The combined thermo-solutal Marangoni convection is suppressed by the variation of surface tension. Especially in the case (d), the concentration distributions in the central r-θ and r-z plane are quite uniform and axisymmetric. However, as Ma_T increases further, the flow pattern becomes unsteady and irregular. The flow stability is disturbed with non-uniform concentration distribution. An obvious track of thermo-solutal Marangoni convection is formed along the free surface of the liquid bridge, and it is quite clear to observe the back flow near the sidewall in the r-z plane. When Ma_T reaches -3500, the pattern becomes much more relaxative compared to that of Ma_T equals -2800, and the flow complexity decreases slightly.

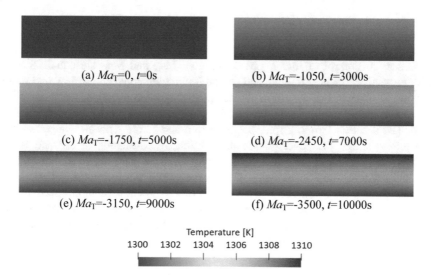

(a) Ma_T=0, t=0s

(b) Ma_T=-1050, t=3000s

(c) Ma_T=-1750, t=5000s

(d) Ma_T=-2450, t=7000s

(e) Ma_T=-3150, t=9000s

(f) Ma_T=-3500, t=10000s

Temperature [K]

1300   1302   1304   1306   1308   1310

Fig. 2. Temperature distribution in the central r-z plane with Ma_C = 1072 and different Ma_T

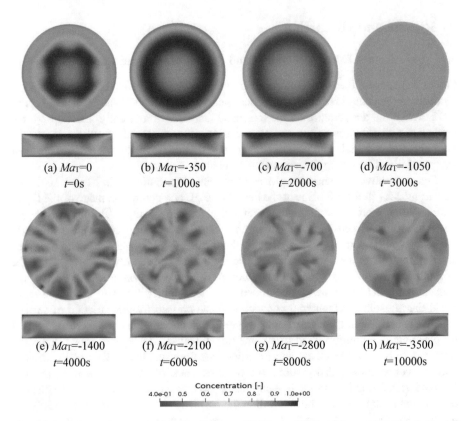

Fig. 3. Snapshots of concentration distribution with $Ma_C = 1072$ and increasing Ma_T in the r-θ plane at the height of $0.5L$ and central r-z plane

3.2 Control of Thermal and Solutal Marangoni Convections by Co-rotation

In the case of control of Marangoni convection, Ma_T and Ma_C are equal 2857 and 1750, respectively, where thermal and solutal Marangoni convections flow in the same direction to observe the control of intensive Marangoni convection. Rotations of top and bottom planes (ω_T and ω_B) were in the same direction, clockwise, with the same speed. The speed is 1 rpm, 2 rpm, 3 rpm, 4 rpm and 5 rpm with $\omega_T = \omega_B$.

Figure 4 shows the concentration distribution by co-rotation of top and bottom planes in the r-θ plane at the height of 0.5L and central r-z plane at 500 s. When there is no plane rotation, the concentration pattern presents a three-dimensional oscillating unsteady flow with azimuthal wave number m = 7. As the rotation speed reaches 1 rpm, there shows a three-dimensional rotating flow in the external region. With the increase of rotation speed continuously, the flow pattern tends to be steady and two-dimensional axisymmetric and gradually extends outwards. Accordingly, in view of the graphs in the central r-z plane, thermal and solutal Marangoni convections near the free surface shrink in the r direction as the rotation speed increases correspondingly.

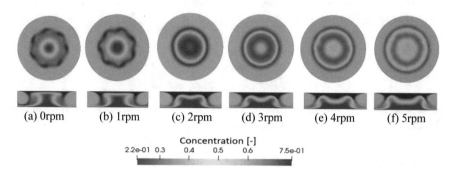

(a) 0rpm   (b) 1rpm   (c) 2rpm   (d) 3rpm   (e) 4rpm   (f) 5rpm

Concentration [-]

2.2e-01 0.3   0.4   0.5   0.6   7.5e-01

Fig. 4. Concentration distribution with $Ma_T = 2857$ and $Ma_C = 1072$ by co-rotation in the r-θ plane at the height of $0.5L$ and central r-z plane at 500 s

Figure 5a indicates the concentration variation by co-rotation at the sampling point $(r, \theta, z) = (0.99a, 0, 0.5L)$. When the rotation speed is 1 rpm, the concentration pattern fluctuates with a regular amplitude and frequency, but varies from time to time. When the rotation speed reaches 2 rpm, the concentration decreases with a slight fluctuation compared to that of 1 rpm. As it increases to 3 rpm, 4 rpm and 5 rpm, the concentration increases with a seemingly steady flow. Figure 5b shows the velocity in the z-direction at the same sampling point by co-rotation. As is shown in the figure, when the rotation speed is under 1 rpm or 2 rpm, there are fluctuations as time goes. With the increase of rotation speed, the velocity decreases with a relatively steady tendency.

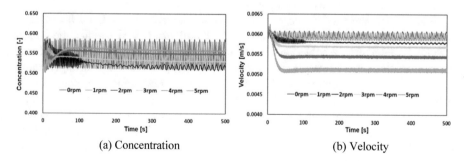

(a) Concentration   (b) Velocity

Fig. 5. Concentration variation and velocity in the z-direction by co-rotation with $Ma_T = 2857$ and $Ma_C = 1072$ at the sampling point of $(0.99a, 0, 0.5L)$

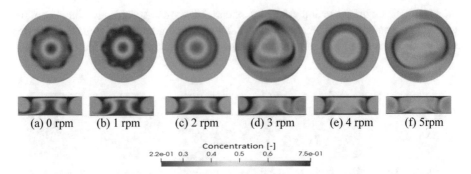

(a) 0 rpm   (b) 1 rpm   (c) 2 rpm   (d) 3 rpm   (e) 4 rpm   (f) 5rpm

Concentration [-]
2.2e-01 0.3   0.4   0.5   0.6   7.5e-01

Fig. 6. Concentration distribution with $Ma_T = 2857$ and $Ma_C = 1750$ by counter-rotation in the r-θ plane at the height of $0.5L$ and central r-z plane at 500 s

3.3 Control of Thermal and Solutal Marangoni Convections by Counter-Rotation

In this case, Ma_T and Ma_C are equal 2857 and 1750, respectively. Top and bottom planes rotate in the opposite direction at the same speed, with counter-clockwise rotation on the top plane and clockwise on the bottom plane. The speed was also set at 1 rpm, 2 rpm, 3 rpm, 4 rpm and 5 rpm with $\omega_T = -\omega_B$.

Figure 6 shows the concentration distribution by counter-rotation of top and bottom planes in the r-θ plane at the height of $0.5L$ and central r-z plane at 500 s. When the rotation speed is 1 rpm, the concentration pattern shows a three-dimensional oscillatory rotating flow, which bulges outwards in a regular frequency, with azimuthal wave number $m = 7$. At the speed of 2 rpm, it is a two-dimensional axisymmetric steady flow. However, as the rotation speed increases further, the flow becomes chaotic with three prominent patterns at the speed of 3 rpm and two prominent parts at 5 rpm. Although, under the condition of 4 rpm, the flow seems axisymmetric and steady, it might be in the transient state under critical rotation speed. Therefore, the concentration pattern is unstable and may become weaker easily under 4 rpm.

Figure 7 presents the concentration variation and velocity in the z-direction by counter-rotation at the sampling point of $(r, \theta, z) = (0.99a, 0, 0.5L)$. When the rotation speed is 1 rpm, there are some fluctuations in both fields of concentration and velocity. As the rotation speed increases to 2 rpm, the flow patterns are seemingly stable, especially in the field of velocity. However, as it increases further, the concentration and velocity vary evidently by a large margin and change disorderly. Accordingly, under the rotation speed of 4 rpm, the flow patterns become more gently, compared to that of 3 rpm and 5 rpm, which also indicates that rotation of 4 rpm makes thermal and solutal Marangoni convections in a transient and relatively unsteady state in accordance with the concentration distribution as Fig. 6 shows.

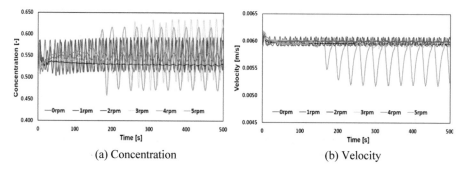

(a) Concentration   (b) Velocity

Fig. 7. Concentration variation and velocity in the z-direction by counter-rotation with $Ma_T = 2857$ and $Ma_C = 1072$ at the sampling point of $(0.99a, 0, 0.5L)$

Although the condition of counter-rotation at 2 rpm can suppress the unsteady flow, it is quite essential to control the rotation speed in an accurate value. Once the rotation speed exceeds or decreases, the flow turns to be unsteady and fluctuated easily. Therefore, it is not propositional to suppress thermal and solutal Marangoni convections by counter-rotation of planes.

4 Conclusion

The present three-dimensional numerical simulation investigated the effects of thermal and solutal Marangoni convections in the opposite direction and its control by co/counter-rotation of planes. The following conclusions were obtained:

1. When thermal and solutal Marangoni convections flow in the opposite direction, the temperature distribution is almost two-dimensional axisymmetric and steady.
2. As $Ma_C > -Ma_T$, the flow shows two-dimensional axisymmetric and steady. As $Ma_C < -Ma_T$, it presents a three-dimensional irregular and unsteady flow.
3. Thermal and solutal Marangoni convections can be effectively controlled by co-rotation of top and bottom planes, which is more than 2 rpm.

References

1. Nishino, K., et al.: Instability of thermocapillary convection in long liquid bridges of high Prandtl number fluids in microgravity. J. Cryst. Growth **420**, 57–63 (2015)
2. Yang, S., et al.: Oscillating characteristic of free surface from stability to instability of thermocapillary convection for high Prandtl number fluids. Int. J. Heat Fluid Flow **61**, 298–308 (2016)
3. Minakuchi, H., et al.: Effect of thermo-solutal Marangoni convection on the azimuthal wave number in a liquid bridge. J. Cryst. Growth **468**, 502–505 (2017)
4. Minakuchi, H., et al.: The relative contributions of thermo-solutal Marangoni convections on flow patterns in a liquid bridge. J. Cryst. Growth **385**, 61–65 (2014)

Comparison of Execution Efficiency of the Use of a Skip List and Simple List in a .NET Application

Igor Košťál[✉]

Faculty of Economic Informatics, University of Economics in Bratislava,
Dolnozemská cesta 1, 852 35 Bratislava, Slovakia
igor.kostal@euba.sk

Abstract. A single level linked list (a simple list) is a dynamic data structure that is used for storing data in applications. However, there are also multi-level linked lists (called skip lists) that are more complicated for creating, but searching for the required data elements in them is more efficient because they allow us to skip to the correct element in them. We have created a C# object-oriented .NET application that uses both lists, a skip and simple list, with structured data in their data elements. By using our C# application we want to compare the execution efficiency of the use of these lists, therefore the same structured data of students is stored in the data elements of its skip and simple list. Our C# application is able to carry out basic operations with these data elements of a skip and simple list, such as searching for students (data elements) according to their points for accommodation, their year of birth, their surname and according to their ISIC in both lists, inserting of a new student (a data element) sorted into both lists etc., and simultaneously it is able to measure the execution times of particular operations. By comparing these times we have examined the execution efficiency of these operations in a skip list and in a simple list. The results and evaluation of this examination are listed in the paper.

Keywords: Skip list · Simple list · Data element · Structured data
.NET application

1 Introduction

It is known that a single level linked list (a simple list) is a dynamic data structure that is used for storing data in applications. One disadvantage of a simple list against an array is that it does not allow direct access to the individual elements. If we want to access a particular item then we have to start at the head and follow the references until we get to that item [1]. A multi-level linked list (a skip list) partially eliminates this disadvantage of the simple list. A skip list is more complicated for creating, but searching for the required data elements in it is more efficient because it allows us to skip to the correct element in this list. There is implementation of skip lists in C programs with simple, unstructured data in their data elements [2, 3]. We have created a C# object-oriented .NET application that uses both lists, a skip and simple list, but with structured data of students in their data elements. Our .NET application is able to

© Springer Nature Switzerland AG 2019
G. Laukaitis (Ed.): INTER-ACADEMIA 2018, LNNS 53, pp. 252–259, 2019.
https://doi.org/10.1007/978-3-319-99834-3_33

carry out search operations and an insertion operation with these data elements of a skip and simple list, and simultaneously it is able to measure the execution times of particular operations. The paper deals with the examination of the execution efficiency of these operations in the skip list and in a simple list using this .NET application. We assume that the use of a skip list in the .NET application should be more effective.

2   A Simple List and a Skip List

A one-way linked list (a simple list) (Fig. 1a) is a set of dynamically allocated elements (called nodes), arranged in such a way that each element contains two items - the data and a reference to the next element. The last node has a reference to NIL (a special node, which terminates the list). We might need to examine every node of the list when searching a simple list. If the list is stored in sorted order and every other node of the list also has a reference to the node four ahead it in the list (a skip list) (Fig. 1b), we have to examine no more than $[n/4] + 2$ nodes (where n is the length of the list) [3]. This data structure could be used for fast searching for required nodes.

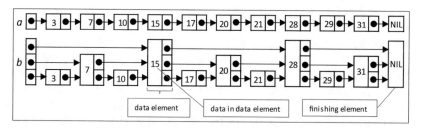

Fig. 1. The simple list (a) and the skip list (b) with simple, unstructured data in their data elements [3]

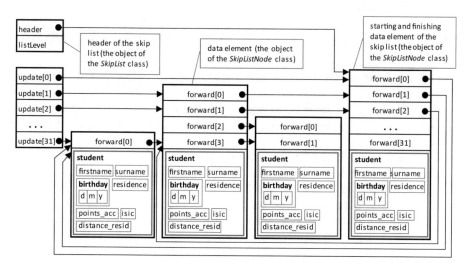

Fig. 2. The internal structure of the skip list that uses our .NET application

We have created a C# object-oriented .NET application that uses both lists, a skip and simple list, but with structured data of students in their data elements (Fig. 2). Using this application, we examine the execution efficiency of the search operations and the insertion operation in the skip list and in the simple list.

3   A .NET Application with Implemented a Simple List and a Skip List

Our .NET application was developed in the C# language in the development environment Microsoft Visual Studio 2013 for the Microsoft .NET Framework version 4 and for the Microsoft operating systems Windows 7, Windows 8.1 and Windows 10. The source code of this application is divided to two parts that are saved into files:

SkipLNodeClasses.cs - the *SkipList* class is defined here. This class includes member methods that are able to perform all search operations such as searching for students (data elements) according to their points for accommodation, their year of birth, their surname and according to their ISIC in a skip list and in a simple list (*find_1st_pts*, *find_1st_pts_SL*, *find_1st_y*, *find_1st_y_SL*, *find_1st_surname*, *find_1st_surname_SL*, *find_1st_isic*, *find_1st_isic_SL*), inserting of a new student (a data element) sorted into both lists (*insert_1st*, *insert_1st_SL*). The *SkipList* class also includes service member methods serving to delete particular data elements or all data elements. Two nested classes *SkipListNode* and *SimpleListNode* are defined within the scope of the *SkipList* class. The *SkipListNode* class object represents one data element (node) of a skip list, and the *SimpleListNode* class object represents one data element data (node) of a simple list. In the *SkipList* class, two structures the *Student* and the *date* are declared, by which the data part of the data element of the skip list and the data element of the simple list is created. From the *SkipLNodeClasses.cs* source file was built (.dll) library into the *SkipListNode.dll* file that is the dynamic run-time component of our .NET application.

The member methods *insert_1st* and *insert_1st_SL* insert one new data element containing data of one student sorted into a skip list and a simple list according to the following criteria:

- first, they place a student into a given list according to points for accommodation points,
- if this student has achieved the same points for accommodation as the other student, then these two students will be sorted according to their distance from the university,
- if this student has achieved points for accommodation and the distance from his/her residence to the university is the same as the other student's, then these two students will be sorted according to their birth dates,
- if this student has achieved the same all three previous parameters as the other student then these two students will be sorted by these methods according to their surnames,

- if this student has achieved the same all four previous parameters as the other student, then these two students will be sorted by these methods according to their first names.

SkipList_WinForm1.cs - the *SkipList_WinForm1* class that is derived from the *System.Windows.Forms.Form* class is defined here. This system class represents a window that makes up an application's user interface. The *SkipList_WinForm1* class includes event handlers of all controls of the user interface of our .NET application and helper methods. The *SkipList_WinForm1.cs* source file was built into the *SkipList_WinForm.exe* file. The *SkipList_WinForm.exe* is a Windows application that uses the dynamic run-time component *SkipListNode.dll*.

Our .NET application has the structured data of students stored in the data elements of a simple and skip list (Fig. 2). This application is able to carry out search operations and an insertion operation with these data elements of a skip and simple list, and simultaneously it is able to measure the execution times of particular operations. By comparing these times we have examined the execution efficiency of these operations in a skip list and in a simple list in an experiment.

4   Experiment, Its Results and Their Brief Analysis

We assume that the use of a skip list with structured data in its data elements in a .NET application is more effective than the use of a simple list with the same structured data in its data elements in the same .NET application. With an increasing number of the same data elements in these two lists, the execution efficiency of particular operations in the skip list should increase compared to the execution efficiency of the same operations in a simple list.

To verify these assumptions, we performed an experiment using our .NET application, which has implemented both lists, a skip and simple list, into its structure variables and methods. The data elements of both lists contain the same structured data of students. Our .NET application is able to carry out by its methods more operations with students (with data elements) stored in the skip list and simple list such as searching for students (data elements) according to their points for accommodation, their year of birth, their surname and according to their ISIC in both lists, inserting of a new student (a data element) sorted into both lists etc., and simultaneously it is able to measure the execution times of particular operations.

The .NET application worked during the experiment with 4 different big sets containing always 2 lists - a skip list and simple list that were used to measure the execution times of all operations. The first set included the skip list with 11 data elements and the simple list also with 11 data elements. Both lists contained the same structured data of students in their data elements. The next 3 sets included the skip lists and the simple lists with 100, 200, and 300 data elements, always with the same structured data of students in the data elements of both lists in one set. Our .NET application is able to create a sorted skip list and a sorted simple list from data of students that is individually inserted into its input text box. This application is also able to create both lists from the data of the students stored in the *.txt* file. This method of creating sorted skip lists and sorted simple lists was used in all measurements of execution times of search operations.

The same search parameter of the searched students was always used during the search operations execution times measurement in the given set of two lists of students. During the insertion operation execution times measurement, the same new student with the following data was inserted into both lists of each set of lists: *Jakub Kadlecik 36018527025247237 9 1 1997 40 Bratislava 20* (first name, surname, ISIC, date of the birth (day month year), points for accommodation, residence, distance between residence and university [km]). The same data format as the data of this new student also has data of students stored in the disc file, e.g. in the above mentioned *students.txt*, from which our .NET application is able to create a sorted skip list and a sorted simple list.

Measurements of the execution times of the particular operations in both lists were performed on the computer with the CPU: *Intel Pentium Processor P6100 (3 MB Cache, 2.00 GHz, 2 Cores, 2 Threads)* and a 4 GB operating memory. The Microsoft Windows 7 Home Premium Service Pack 1, 64 bit operating system and the Microsoft .NET Framework 4 were installed on this computer. Our .NET application was developed to this target framework and therefore it must be installed on the experimental computer.

One of the outputs of our .NET application displayed in its text box after searching for students with 61 points for accommodation in a skip list with 100 data elements, is shown in Fig. 3. The same output writes the .NET application to the *LogFile.txt* disc file. The .NET application writes the results of other search operations to the same *LogFile.txt* file, too.

[ISIC surname first name	(points for acc.) residence (dist. to univ.) date of the birth]
36099605837843210 Diko Adrian	(61) Dunajska-Streda (101) 1996-04-13
36015378227465325 Sinaj Jan	(61) Dunajska-Streda (101) 1996-04-12
36104458148066224 Marbetik Janny	(61) Bratislava (21) 1999-02-17
Search time: 00:00:00.0008571, 0,8571 ms	

Fig. 3. The output of our .NET application displayed in its text box after searching for students with 61 points for accommodation in a skip list with 100 data elements

The execution times of all search operations and the insertion operation performed by our .NET application on all four sets of two lists are shown in Figs. 4 and 5.

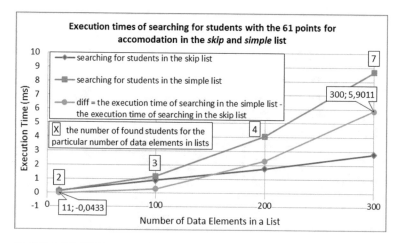

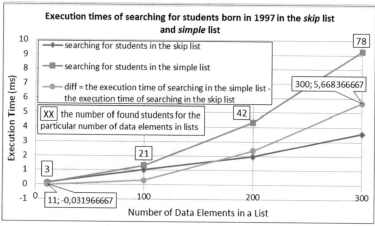

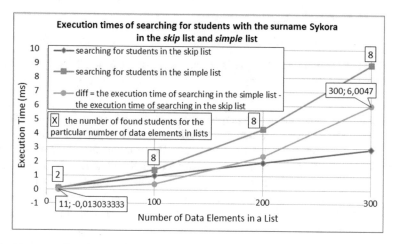

Fig. 4. The execution times of all search operations performed by our .NET application in the skip list and simple list

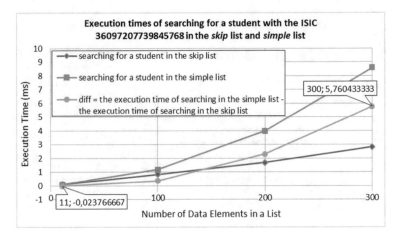

Fig. 4. (*continued*)

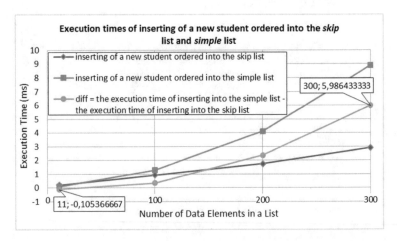

Fig. 5. The execution times of the insertion operation performed by our .NET application in the skip list and simple list

Brief Results Analysis. From the graphs depicting the dependency of the execution times of all search operations and the insertion operation performed by our .NET application in the skip list and the simple list on the numbers of data element in these lists, it is obvious that the operations performed in the skip list have significantly shorter execution times at higher numbers of data element compared to the operations performed in the simple list. A significant difference between these execution times can be seen at the number of data elements 100. This difference increases with the increasing number of data elements in both lists. Based on this, it is obvious that the use of the skip list with structured data in its data elements in the .NET application for more than 100 data elements is more efficient than the use of the simple list with the same structured data in its data elements in such an application.

5   Conclusion

From the results of the comparison of the execution times of the search operations and the insertion operation performed by our .NET application in the skip list and execution times of the same operations performed by the same application but in a simple list, it is obvious that the use of the skip list in this application is with a bigger number of data elements containing structured data, significantly more efficient. From these comparison results, we can also say that for small numbers of data elements containing structured data, e.g. for 11 data elements, the execution efficiency of the use of the skip list and simple list in the .NET application is approximately the same. However, the cost of implementing a skip list into a .NET application is higher than the cost of implementing a simple list. If our .NET application searches for, inserts data or performs other operations with a small number of data elements containing structured data, then it is probably better to use a simple list in such application. If we use a big number of such data elements, 100 or more, in a .NET application, it is a good choice to use a skip list in such application.

References

1. Linked Lists. http://www.cs.cmu.edu/~adamchik/15-121/lectures/Linked%20Lists/linked%20lists.html. Accessed 19 June 2018
2. Niemann, T.: Sorting and searching algorithms (1999). www.epaperpress.com
3. Pugh, W.: Skip lists: a probabilistic alternative to balanced trees. Commun. ACM **33**(6), 668–676 (1990)

An Application of Nash Equilibrium to an Experimental Setting: The Real Meaning of the Sacrifice Move in Board Games

Naomichi Ikarashi[1], Raiya Yamamoto[2], and Valerie A. Wilkinson[1(✉)]

[1] Faculty of Informatics, Shizuoka University, 3-5-1 Johoku, Naka-ku,
Hamamatsu, Japan
vwilk@inf.shzuoka.ac.jp
[2] Faculty of Engineering, Sanyou-Onoda City University, 1-1-1 Daigaku-dori,
Sanyo-Onoda, Japan

Abstract. This is a case study of one student's research about the Nash equilibrium in the context of board games. First set the stage, i.e. describe the educational environment. In 1995 Faculty of Informatics was a new faculty. The faculty itself was a "**simulation**," establishing a viable institution of higher learning. An irregular procedure made it possible for a **B**achelor's degree **4**th year (B4) in computer science (CS) student to choose the General Systems Theory (GST) Lab's (LAB) Event Planning and Game Theory (GT) lab for graduation research. The lab has keywords GST, experiential learning, GT, requisite variety, etc. Educational environment impacts strategic learning (Young 2004), at the level of student, faculty, staff, and university. The third player in this account is a **M**aster of **S**cience (MS) candidate who was LAB's **T**eaching **A**ssistant (TA).

3 players/stakeholders hold different roles in this account: LAB's job is to provide praxis of **requisite variety** and a layered **educational environment** for Science, Technology, Engineering, and Mathematics (STEM) students. This story is a learning story with a template. B4 required an environment for experiments with main interest in the **board game** shogi. How B4 linked the "sacrifice move" to Nash equilibrium is the goal of this narrative. TA provided crucial insight at the right time.

Keywords: Board-games · Sacrifice move · Nash equilibrium
Educational environment

1 Introduction

Why did Mr. Igarashi (B4) choose GST LAB? In 2013 we had a chance to guide the research of a CS student who chose to work in the General Systems Theory (GST) LAB. The Lab's focus is Event Planning and Game Theory (GT). LAB was not the most likely candidate, emphasizing neither technology nor mathematics.

B4 chose the lab and decided to study the Nashi equilibrium in the context of the board game shogi. That is the center and main point of this research. LAB had several

© Springer Nature Switzerland AG 2019
G. Laukaitis (Ed.): INTER-ACADEMIA 2018, LNNS 53, pp. 260–265, 2019.
https://doi.org/10.1007/978-3-319-99834-3_34

students during tenure at Faculty of Informatics; B4 uniquely chose Game Theory, Nash equilibrium, and board games.

The method of this study must be a case study, to frame and reveal the accomplishment of this buddying game theorist.

"A case study is expected to catch the complexity of a single case." (Stake 1995)

The technique of the case study is to reveal and situate the focus of attention in an environment, in this case, the structured educational environment of GST LAB. Thick description and triangulation of the case study offers various perspectives, and reflection with which to gain a glimpse of what B4 was able to achieve. TA had chosen to build up language skills by using an available language resource. His presence and point of view (POV) altered the terrain.

The three players in this narrative each hold developmental stakes and engagement in this situation. While the student is the proper center of attention for this article, the Lab setting is necessary to frame the case.

1.1   The Lab: "Community Space"

For an educator tasked with adapting to the needs of students in Informatics GST provides a borderless, observable, adaptive, and easily validated learning environment. In the first place, LAB has praxis, i.e. the implementation of dialectic anywhere, on any day, in any situation. Next, GST LAB uses Gregory Bateson as the exemplar of the whole, living system mindset. Using dialectic as a tool can be initiated with a contrasting pair of concepts, i.e. opposites, let's start with Gregory Bateson's "rigor and imagination:"

"...we shall know a little more by dint of rigor and imagination, the two great contraries of mental process, either of which by itself is lethal. Rigor alone is paralytic death, but imagination alone is insanity." (Bateson 1979)

In a sense, GT illuminates the activity system of GST. GT probes the interfaces of living systems as they come into alignment, harmony, balance, and equilibrium. Our faculty is home to social informatics aka Information Arts (IA) and CS. The interactive learning problem in a faculty which is home to IA and CS majors is itself a game: can the social sciences live together with hard sciences in harmony?

H. Peyton Young notes, "In a social system, individuals' intentions are in equilibrium when no one wants to deviate from his intended behavior given the intention of others." (Peyton 2004). LAB gives us an opportunity to test this theory.

1.2   The TA Finds a Home Away from Home

TA came to the GST LAB in search of a "3rd Place" (Oldenburg 1989). In Ray Oldenburg's words, "The development of an informal public life depends on people finding and enjoying one another outside the cash nexus." In the case of the academic arena, this maxim would be interpreted as "outside of the credit nexus." Students could come to Community Space and have something to eat or drink, spend a while relaxing, or join in some discussion or activity. Since the LAB is an English environment, it is a good place to hang out and build language skills naturally.

The LAB and TA dyad dynamic generated a low frequency resonance which signals to visitors that the space is comfortably inhabited by life-forms.

1.3   He Came, He Saw, He Took His Time About It

B4 found the option of choosing GST LAB which only became possible in 2011, by dint of a codicil in the lab selection protocol: "A teacher can only promote a lab if a student requests it." B4 chose the lab and its denizens did what they could to welcome him. Then, it was up to LAB and B4 to forge the dyadic teacher-student relationship by which the student achieves momentum to enter "research mode."

"The attainment of equilibrium requires a disequilibrium process." (Arrow 1986)

2   B4 Preliminary Diagnostic Interview and Training

B4 reviewed his various interests in LAB sessions and considered various research activities. For a GT activity, LAB hosts a "Community Space: Learning with Games at the University Techno-Festa." We discussed setting up an experiment in game playing. Also, there were GT books in Japanese and English to read.

In case study research, "researcher bias" can affect the interpretation. In particular, case study depends on background and setting (social interpretations of the environment) to present the narrative from a particular perspective. All of this material occurred six or seven years ago which may further affect the choices made for interpretation. LAB background, B4 foreground? B4 is the main player. What did he do?

The first experiment was the Community Space experiment at the University Festival. For the experiment, we proposed that B4 set up a Shogi table in the Community Game Space to play with any comers. We got middle-school students and retired citizens. B4 won; they were satisfied. Thus the game of shogi became a topic the the GST Lab. Then a few of the regulars challenged B4; he won.

With the experiment B4 established himself in the lab. From being an occasional presence, B4 became a "force to be reckoned with" in his own terms. He accomplished this adroit maneuver by himself, with his own resources. However, years later, LAB narrative inferred that the community space at the festival, as well as the experiment design, conferred the mindset.

The second game experiment: "Shoot the Moon" in Hearts. LAB utilizes card games to create a dynamic, awake, and harmonious classroom atmosphere. The first day of fresher's class requires an ice-breaking technique. What better than to institute a lesson in the game of Hearts. B4 was encouraged to set up a Hearts experiment series. The experimental question: could participants learn to "shoot the moon."

The program for conducting the experiment was set up with the hope that the experiment would continue long enough to make a difference. We (1) ordered box lunches (2) set the schedule from 17:45–19:15.

The participants were asked to submit profile surveys. The profile surveys with dates and respondents (i.e. May 21, 5; May 28, 7; June 11, 4; June 25, 8; July 9) are the first half of the "hard" data. The second half of the "hard" data are the participants' score sheets which LAB preserves. Note: The score sheets were influential in catalyzing this research.

3   B4 Studies Nash Equilibrium; Conceives the Analogous "Sacrifice Move"

From B4's thesis: "In my laboratory, we mainly focus our study in two applied fields, community event management and applied game theory. I have chosen to work with Game Theory. ... I will focus on the games of strategy, especially zero-sum games. A zero-sum game is a mathematical representation of a situation in which a participant's gain (or loss) of utility is exactly balanced by the losses (or gains) of the utility of the other participant. This class of games contains classic games such as Igo, Chess, and Shogi. ... I considered whether the Nash Equilibrium might be useful in predicting moves in initial or advanced configurations of these games." (Igarashi 2014).

As B4 understood the Nash equilibrium, "Nash players cooperatively accept rational profits or advances that are small. To make progress using Nash equilibrium would be to consistently take every small advantage." He continues, "Philosophically and ethically this will show up as acceptance and patience. Actually, game players want to win. Using Nash equilibrium they can win, if they accept small gains." p. 8.

B4 goes on to explain the structure of this strategy. "In most of these war games, there are patterns for making the opponent suffer by playing a sacrifice move. If you play a move in a complicated state of game there is a possibility that you would seem to be at a disadvantage, but the situation will change." p. 10.

B4 illuminates the principle that the sacrifice moves to win seemingly are important in the zero-sum games. "To sacrifice a piece at first looks like a loss, but may put one in a more advantageous position." With this generalization, B4 began his research by locating "the sacrifice move," i.e. to give up a piece for a strategic advantage, in Shogi and Chess. He found applications in many other games.

B4 goes on to translate the Nash equilibrium as a "sacrifice move" in human-relation contexts such as trade-offs and patience. The demonstration of B4's thesis requires illustrations. He showed specific moves in specific games with graphs, diagrams, and matrices. It will take a different framing to give B4's thesis a proper presentation. When B4 began interpreting and applying the idea to human contexts, he engaged with the LAB in a new way. The unrewarding matrices showing very small gains began to appear more interesting and pertinent.

4   Reading the Score Sheets with New Eyes

One day, about five years after B4's thesis, after he graduated and went to work, the director of LAB idly flipped through the score sheets that B4 had amassed during the experiment series in 2013. Something was disturbing about the information, something that wasn't there. There were very few "shoot the moon" scores and none of the games were completed.

Realization sank in, that LAB had provided box lunches for a group of students to play Hearts "in good faith," assuming that they would try to win, but they had not been interested in winning, or even finishing a game, thereby invalidating the premise of the experiment. This "travesty" went on for months. "What a sap I've been!" was the lament, and indeed, the director thought that the experimental series had not only been a failure, but an ill-conceived flop from the beginning.

5   A Reinterpretation of the Nash Equilibrium

TA had been a Third Space Community (3SC) member, involved in LAB events as a stake-holder and advisor for many years, as he completed his doctoral work. TA saw the records and heard the director's complaint. He argued that those participants had not shown bad faith by coming to play the game. It was, he said, an example of the Nash Equilibrium, just as B4 described it. The students came because they were invited, they did as they were asked, just exactly for as long as they were asked to.

The director complains, "It looks like they came for the food, they didn't want to be there, and only pretended to be interested." On many levels, the director felt "played" by the students.

Many observations ratified TA's interpretation. The experiment should be taken as a success. The students showed up and did as they were asked. Whether they were kind, respectful, and glad of the opportunity or not, does not matter. They did show up and gave up doing other things to be there. With that, whole new dimensions of this Japanese courteous world became visible in LAB for the first time. Also, those boring matrix representations of Nash equilibria began to make sense in daily life.

6   Discussion

B4 referred to the practice in GST LAB in the introduction to his thesis. "We hold Game Night every Tuesday to encourage participants to think about Game Theory. We play mainly Hearts because it's a very understandable game to learn Game Theory." p. 4.

Those words express the purpose of LAB. Then, is B4 just saying that to "be nice," or does he think so? This element of doubt, leaving LAB floundering in bewildered humiliation is absolutely OK. Some might say, "Thinks too much!"

The experimental series did not fail. Based on the assumptions of the experimental design, it did not prove what LAB wanted to prove or accomplish what LAB wanted to accomplish. It turns out, of course, that experiments with groups are highly susceptible to various interpretations, by any of the members.

The case study is the best way that we could present these findings from the research of B4, who is a true game theorist and mathematician. It's as though LAB provided a nest for a very different kind of bird than expected. Yet how fortunate that one such student came.

7 Conclusion

LAB set about constructing Community Space and Game Night many years before one student made his way to the GST lab to study Game Theory. In presenting the environment of the individual case, LAB was uneasily aware that Researcher Bias would affect the findings.

Michael Polanyi took up this very problem in The tacit dimension (Polanyi 1967). "Suppose that tacit thought forms an indispensable part of all knowledge; then the ideal of eliminating all personal elements of knowledge would, in effect, aim at the destruction of all knowledge."

The GST framework makes it possible to work with an autonomous system embedded in a larger autonomous system, to be observed and assessed by an external, though connected, autonomous system.

Who will tell what is revealed in this case study. Is the vision of B4 visible? Most of his framing and context had to be trimmed to fit the space. The way of organizing his study and lab was part of the case study, involving LAB and TA as players. By editing his work to present it, LAB found it necessary to reorganize it. What is not visible is the 5 years it took to conceive of a possible framing for B4's work, from LAB's limited perspective.

With this, LAB thanks TA for kindly interpreting the meaning at three levels: the research, the experiment series, the lab dynamics, and LAB's understanding. Nash equilibrium is subtle and deep.

References

Young, H.P.: Strategic Learning and Its Limits, p. 1. Oxford UP, Oxford (2004)

Stake, R.E.: The Art of Case Study Research, p. xi. Sage Publications, Thousand Oaks (1995)

Bateson, G.: Mind and Nature: A Necessary Unity, p. 219. E.P. Dutton, New York (1979)

Arrow, K.: Part 2: The behavioral foundations of economic theory. J. Bus. **59**(4), S385–S399 (1986)

Igarashi, N.: The variations of the game theory by the sacrifice moves. Graduation thesis, Shizuoka University (2014)

Polanyi, M.: The Tacit Dimension, p. 20. Doubleday, New York (1967)

A Hybrid Machine Learning Approach for Daily Prediction of Solar Radiation

Mehrnoosh Torabi[1], Amir Mosavi[2,3,4(✉)] ⓘ, Pinar Ozturk[4],
Annamaria Varkonyi-Koczy[3,5], and Vajda Istvan[3]

[1] Hormozgan Regional Electric Co, Bandarabbas, Iran
[2] Institute of Advanced Studies Koszeg, iASK, Kőszeg, Hungary
[3] Institute of Automation, Kando Kalman, Faculty of Electrical Engineering,
Obuda University, Budapest 1431, Hungary
amir.mosavi@kvk.uni-obuda.hu
[4] Department of Computer Science, Norwegian University of Science
and Technology, Trondheim, Norway
amir.mosavi@ntnu.no
[5] Department of Mathematics and Informatics, J. Selye University,
Komarno, Slovakia

Abstract. In this paper, we present a Cluster-Based Approach (CBA) that utilizes the support vector machine (SVM) and an artificial neural network (ANN) to estimate and predict the daily horizontal global solar radiation. In the proposed CBA-ANN-SVM approach, we first conduct clustering analysis and divided the global solar radiation data into clusters, according to the calendar months. Our approach aims at maximizing the homogeneity of data within the clusters, and the heterogeneity between the clusters. The proposed CBA-ANN-SVM approach is validated and the precision is compared with ANN and SVM techniques. The mean absolute percentage error (MAPE) for the proposed approach was reported lower than those of ANN and SVM.

Keywords: Global solar radiation · Prediction
Support vector machine (SVM) · Machine learning
Artificial neural networks (ANN)

1 Introduction

Renewable energy systems aim at satisfying the ever increasing energy demands in a sustainable manner through reducing the greenhouse emissions and climate change risk reduction [1, 2]. Among the renewable energies, the solar is generally considered as the most promising sources, partly due to its availability [3, 4]. As a consequence, we are seeing an increase in solar energy technologies. However, the capability to maximize the utilization and efficiency of solar energy remains a difficult task, partly due to challenges in the collecting and accurate analyzing of the solar radiation data. Nevertheless, the solar energy projects can highly benefit from a reliable solar radiation information. In fact, the global solar radiation is a highly relevant parameter in monitoring, simulating, prediction, and sizing of solar energy technologies [5–9]. Thus, it is

© Springer Nature Switzerland AG 2019
G. Laukaitis (Ed.): INTER-ACADEMIA 2018, LNNS 53, pp. 266–274, 2019.
https://doi.org/10.1007/978-3-319-99834-3_35

essential to be able to accurately predict the solar radiation using proper techniques even at the absence of adequate data.

Several data mining techniques have been employed in business and medical sciences [10], and in recent times, the focus has been on exploring approaches to determining patterns in data set that can be used for description and prediction. Data mining is considered as an inductive machine learning (ML) technique, where the past data set is utilized for training and learning the model of interest. This learning is representing via determining the relationships among the variables and extracting meaningful patterns. The objective of data mining is to use these meaningful patterns for the purpose of accurate prediction [11, 17]. Artificial neural network (ANN) and support vector machines (SVMs), two well-known data mining techniques, have been successfully used to estimate global solar radiation. For example, Mubiru and Banda [12] used ANN technique for estimation of monthly mean daily global solar irradiation at several locations in Uganda. Jiang [13] proposed an ANN model to estimate monthly mean daily global solar radiation in different cities of China. The evaluation of their model shows better precision than the empirical models examined in the paper. Najafi et al. [14] developed a coupled ANN algorithm to predict daily solar radiation in a number of cities in Iran. It was found that the proposed algorithm achieves a better performance than the Angström-Prescott model. Mathioulakis et al. [15] applied an advanced ANN technique in the daily prediction. In their novel work, a number of different sets of input parameters has been used. They further propose ANN as an effective method to predict the solar radiation for a global estimation. Azeez [16] studied the monthly prediction through using maximum ambient temperature, Sunshine duration, and relative humidity as the required input parameters. In addition, Mosavi et al. [17] reviewed similar methods of prediction. In another study, Chen et al. [18] evaluated the usage of SVMs for predicting the monthly mean based on the site's minimum and maximum temperature employing different functions of SVM with promising results. Furthermore, Chen et al. [19] proposed a number of duration-based SVM algorithms which showed superior results. Mosavi and Varkonyi [20] also utilized SVMs to predict solar radiation considering the ambient temperature. Chen and Li [21] assessed the performance of 20 SVM for estimation and reported that using SVM-based models could result in better accuracy compared to ANN models.

Guermoui et al. [22] evaluated the utility of two support vector regression (SVR) models, based on the radial basis function and the polynomial basis function, for prediction of monthly mean daily global solar radiation. Their funding's indicated SVR based on the polynomial basis function have better accuracy over SVR based on the radial basis function [23, 27, 28]. A number of authors have also attempted to achieve better accuracy in estimating solar radiation using the hybrid approaches. For example, Wu et al. [24] integrated the time delay neural network (TDNN) with autoregressive and moving average (ARMA) algorithm to predict hourly solar radiation. The hybrid model provides a higher capability compared to either TDNN model or the ARMA model alone. Similarly, Moeini et al. [25] proposed a hybrid approach of fuzzy and hidden Markov models to effectively predict the solar irradiation. Their results demonstrated that the predictions of the proposed model are close to the training data set. Halabi et al. [29] developed a hybrid approach by integrating simulated annealing (SA) and genetic programming (GP). The results of their sensitivity analysis showed that the suggested model provide accurate predictions. Guermoui et al. [22] compared the precision of a hybrid SVM model with ANN and GP. As an alternative, we propose a new concept to estimate global solar radiation on a horizontal surface, using a cluster-based approach (CBA). Our CBA utilizes both ANN and SVM approaches to accurately estimate daily global solar radiation, and this new approach is hereafter referred to as CBA-ANN-SVM. This hybrid approach enjoys the benefits offered by both ANN and SVM as well as those of the clustering technique. Clustering analysis classifies the global solar radiation data into various clusters. This allows us to maximize the homogeneity of data within the clusters as well as maximizing the heterogeneity between the clusters. To test the validity of the proposed method, we use measured data over a period of 10 years, including different meteorological variables and the horizontal radiation, from Kerman region in Iran. We then compared the performance of the proposed CBA-ANN-SVM method against those using ANN and SVM techniques.

2   Description of Data Collection

The city of Kerman located is the capital of Kerman province in Iran is used as the case study in this paper. This studied site is located between 32°N and 25°55/N and also between 53°26/E and 59°29/E. This location is in the sunniest spot of the region with the sea level elevation of 1,756 m and the location of 30°29/N and 57°06/E. The region has a dry and moderate climate. According to the long-term measured data, the monthly average air temperature varies from 4.6 °C to 26.8 °C and the yearly average is 15.9 °C. The monthly average relative humidity varies between 19% and 53% with the annual average of 32%. The data set includes 10 years daily sampled data, consisting of the horizontal global solar radiation (H), sunshine duration (n), maximum and minimum air temperature (Tmax and Tmin) for the period of December 1994 to January 2005. In this study, to filter the data sets and reduce the abnormalities and inconsistencies in the values the concept of daily clearness index (K_t) was used. For this aim, we compute K_t and determine and omitte the values of the out of range of $0.015 < K_t < 1$ [26, 30]. K_t is defined as the ratio of horizontal global solar radiation (H) to the radiation on a horizontal surface (Ho). To model the horizontal global solar radiation via proposed method, the parameters of n, T_{max}, T_{min}, Ho and maximum possible sunshine duration (N) are considered as inputs. Furthermore, the values of N and H_o were computationally modeled utilizing the equations.

Table 1. Pearson correlation coefficient between the global solar radiation and input variables.

**. Correlation is significant at the 0.01 level (2-tailed).		n**	N**	T_{min}**	T_{max}**	H_o**
Pearson correlation coefficient	H	0.716	0.825	0.646	0.764	0.822

To identify the influence of considered parameters on accurate prediction of global solar radiation, the Pearson correlation coefficient between the dependent parameter (output) and independent parameters (inputs) were calculated using SPSS software.

Table 1 presents these achieved Pearson correlation coefficient. According to the Table 1, it is noticed that all considered inputs have favorable correlations with global solar radiation. However, the highest correlation is achieved for maximum possible sunshine hours (N) while the lowest correlation is obtained for minimum air temperature (T_{min}). As one of the most effective graphical methods to determine the correlation, pattern or trend between two parameters is the scatter plot, to illustrate the correlations attained between global solar radiation and the considered inputs parameters their scatter plots are depicted. The scatter plots between H and the inputs n, N, T_{min}, T_{max} and H_o are shown in Fig. 1(a–d), respectively.

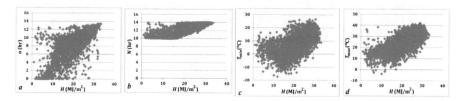

Fig. 1. Scatter plots of horizontal global solar radiation and the considered input parameters

3 Modeling

In order to build the models, the Clementine software version 12.0 has been utilized. Three different methods including the SVM, ANN and the hybrid cluster based method that uses ANN and SVM (CBA-ANN-SVM) have been developed and used for this research work. In the following, all developed models are explained, and then the best model with least estimation error is determined.

3.1 Implemented Model Using SVM Approach

SVM is one of the new and well-known ML approaches. It is capable to perform favorably even when the data samples are limited or they are non-linear and also the dataset is high-dimensional or there exist local minima. SVM is also capable of high generalization. Figure 2 illustrates the implemented model based upon the SVM approach. For modeling, initially, the used data sets are brought to Clementine. Source node that has been named "Imported Data", reads in data from external source (dataset that we have preprocessed) into Clementine. A "Partition" node is utilized to split the data into separate subsets or samples for training and evaluation stages of model building. For this study, 50% of the data sets were used for the training purpose and 50% of data sets were utilized for the testing purpose. Partition node has "random seed" option. By this option, we can ensure different samples (by selecting another subset of data records) will be generated each time the node is executed. By "Type" node, we tell Modeling node ("SVM" node) whether fields will be predictor fields or predicted fields. This node also describes data type (string, integer, real, date, time, or timestamp) in a given field. "SVM" node is a Modeling node. This sequence of operations is known as a data stream. When the stream is executed and model is built, the model nugget ("SVM-Energy") is created and added to the Models palette in the upper right corner of the application window. In accordance with Clementine software, to see modeling result we have to add the model nugget to the stream and attach the model nugget to the "Type" node, at the same point as the Modeling node. "Analysis" node helps to determine whether the model is acceptably accurate. Building the SVM model requires a trade-off between maximizing the margins and the minimizing learning error. The Clementine software has a regularization parameter "c", which is used to regulate this trade-off. Increasing c leads to higher classification accuracy (reduced regression error) but it may also cause overfitting. In this study, three different kernel functions of linear, polynomial, and sigmoid are tested. After building each model, its performance to estimate global solar radiation was evaluated by calculating the mean absolute percentage error (MAPE) and standard deviation (SD). The MAPE is obtained by:

$$MAPE = \frac{1}{N} \sum_{i=1}^{N} \left| \frac{H_{esti}^i - H_{maes}^i}{H_{meas}^i} \right| \times 100 \qquad (1)$$

Where H_{esti}^i and H_{meas}^i are the i_{th} predicted the global solar radiation values, respectively, and N represents the total number of data samples. In order to develop the final SVM model with the lowest MAPE, a polynomial function with adjustment parameter of 8 and gamma parameter of 2.5 was used. Table 2 shows the attained MAPE and SD values for prediction of global solar radiation employing the proposed SVM model. The significance of each considered input element to predict global solar radiation based through the proposed SVM is shown in Fig. 2. According to the Fig. 2, it is noticed that, T_{min} has a little importance on estimation of global solar radiation using SVM model while the highest importance belongs to the N.

3.2   Implemented Model Using ANN Approach

The second method employed to predict global solar radiation is advanced on the basis of ANN technique. The implemented model on the basis of ANN is also shown in Fig. 2. Similar to the SVM method, in the beginning, the used data sets are brought to Clementine. The "Partition" node is used to divide the data into two subsets for training and evaluation stages of model building. After building the model, the "Analysis" node is used to determine that whether there is any overfitting. Considering the supervised ANN, every single learning phase is named a cycle. These cycles continue till the networks' weight get stable. The parameter "Persistence" is set equal to 400 in this model which means that if till 400 cycles the error would remain constant then the model has become stable. For various settings, the global solar radiation modeling was conducted and subsequently the MAPE and SD values were computed. The final and best model was built with one layer of input, two hidden layers and one output layer. The achieved MAPE and SD values using the best ANN model developed is presented in Table 2. It is observed that n is the most relevant element whereas T_{min} and T_{max} of which influences on estimation are close to each other have the least significance.

3.3   Implemented Model Based on Clustering (CBA-ANN-SVM Approach)

Another model developed in this research work is based upon clustering. The goal is to verify the strength of clustering for global solar radiation estimation. The architecture of hybrid cluster based model is as follow: Step 1: Clustering, Step 2: Modeling for each cluster. One of the important points regarding the clustering is determining the number of clusters. The two step algorithm has the advantage which makes it possible to specify the number of clusters manually. Also, the algorithm can calculate the number of clusters automatically. In fact, there is no need to initial choice of the number of clusters. In addition, the algorithm is not sensitive to outliers' data, although in this study the outliers were omitted from data sets using solar data cleaning process. Thus, the two step algorithm has been utilized in this study. To analyze the rules on clusters, the decision tree and c5.0 algorithm have been used as presented in Fig. 2.

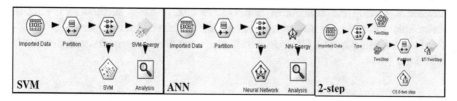

Fig. 2. Model implementation using SVM, ANN and the 2-step algorithm for clustering.

The clustering was performed on the basis of considering different variables such as: (1) H and month, (2) H, number of month and n as well as (3) H, month and number of days. For all their cases, the data sets were clustered to 12 clusters based on the number of months. Thus, the number of month is the influential variable in clustering. According to the analysis conducted using c5.0 algorithm, the governing rules on the clusters are presented. Thus, in the first step, based on the variable of month and using the unsupervised learning method, the inputs are clustered and divided to a series of sub-sets which have the similar features (homogeneous groups). In the next step, the estimations are conducted separately in each clusters using one of the techniques of ANN and SVM, considered as supervised learning. Figure 2 offers a graphical representation of data and distribution fields (H and number of month) between the clusters. It shows that the significance of variables is equal to 1 which indicates the high importance of these two variables in clustering. After clustering, modeling was performed separately on each cluster. For each cluster, the SVM and ANN methods were used. The data sets were divided into two subsets for training and testing by Partition node. To obtain the final error of models, the results of the clusters were combined together separately. Figure 2 illustrates the implemented model on the basis of hybrid cluster based method. In the Table 2, the utilized models as well as the obtained values MAPE and SD for the hybrid cluster based approach are presented for each cluster.

Table 2. The obtained MAPE and SD for the SVM, ANN and CBA-ANN-SVM models.

Model	MAPE	SD
SVM	1.565	2.806
ANN	1.603	2.735
CBA-ANN-SVM	1.342	2.256

3.4   Performance Comparisons

Table 2 presents the comparisons between the performances of all three models based on obtained MAPE and SD values.

In the hybrid cluster based approach (CBA-ANN-SVM), it is found that number of months is an important factor in clustering. In fact, during the data clustering, the data sets are assigned in the target cluster based upon the number of months. Afterwards for estimation of horizontal global solar radiation, the proposed model utilizes the target cluster according to the number of months. The results offered in Table 2 is the verification regarding the benefits of utilizing the cluster based method to predict the global solar radiation. As the lowest error values is achieved for the hybrid CBA-ANN-SVM model, this model is introduced as the superior one for estimation of global solar radiation.

4   Conclusions

In this study, a Cluster-Based Approach (CBA) was introduced to estimate daily global solar radiation on a horizontal surface. For this aim, the clustering paradigm along with ANN and SVM techniques were utilized in our proposed hybrid approach (CBA-ANN-SVM). To demonstrate the practicality of CBA-ANN-SVM, we evaluated the approach using 10 years of measured data sets from an Iranian city located in a sunny part of the country. The measured sunshine hours, calculated the maximum amount of the possible sunshine hours, maximum and minimum air temperatures, and extraterritorial solar radiation were used as inputs for the prediction of global the solar radiation. Clustering was performed to categorize the global solar radiation data into the clusters. It was found that number of months is a significant parameter in clustering. To achieve this, the clustering was performed according to the month of the year, so that the data sets could be clustered into 12 clusters based on the month. This allowed us to maximize the homogeneity of data within the clusters and the heterogeneity between the clusters. Our evaluation of the CBA-ANN-SVM approach indicated that this approach resulted in a higher accuracy compared to using ANN and SVM techniques. For example, the MAPE using our approach is 1.342%, as compared to 1.603% and 1.565% using ANN and SVM, respectively.

Acknowledgment. This work has partially been sponsored by the Hungarian National Scientific Fund under contract OTKA 129374 and the Research & Development Operational Program for the project "Modernization and Improvement of Technical Infrastructure for Research and Development of J. Selye University in the Fields of Nanotechnology and Intelligent Space",

ITMS 26210120042, co-funded by the European Regional Development Fund. Dr. Mosavi contributed in this research during the tenure of an ERCIM Alain Bensoussan Fellowship Programme. The support and research infrastructure of Institute of Advanced Studies Koszeg, iASK, is acknowledged.

References

1. Hernandez, R.: Environmental impacts of utility-scale solar energy. Renew. Sustain. Energy Rev. **29**, 766–779 (2014)
2. Hosseini, E.: A review on green energy potentials in Iran. Renew. Sustain. Energy Rev. **27**, 533–545 (2013)
3. Torabi, M., et al.: A Hybrid Clustering and Classification Technique for Forecasting Short-Term Energy Consumption, Environmental Progress & Sustainable Energy. Wiley, Hoboken (2018)
4. Mekhilef, S.: A review on solar energy use in industries. Renew. Sustain. Energy Rev. **15**, 1777–1790 (2011)
5. Imani, M.H.: Strategic behavior of retailers for risk reduction and profit increment via distributed generators and demand response programs. Energies **11**(6), 1–24 (2018)
6. Rusen, S.: Estimation of daily global solar irradiation by coupling ground measurements of bright sunshine hours to satellite imagery. Energy **58**, 417–425 (2013)
7. Darvishzadeh, A.: Modeling the strain impact on refractive index and optical transmission rate. Physica B: Condens. Matter **543**, 14–17 (2018)
8. Ulgen, K., Hepbasli, A.: Diffuse solar radiation estimation models for Turkey's big cities. Energy Convers. Manag. **50**, 149–156 (2009)
9. Karakoti, I., Pande, B., Pandey, K.: Evaluation of different diffuse radiation models for Indian stations. Renew. Sustain. Energy Rev. **15**, 2378–2384 (2011)
10. Mosavi, A.: The large scale system of multiple criteria decision making. Large Scale Complex Syst. Theory Appl. **9**(1), 354–359 (2010)
11. Vargas, R., Mosavi, A., Ruiz, L.: Deep learning: a review. In: Advances in Intelligent Systems and Computing (2017)
12. Mubiru, J.: Estimation of monthly average daily global solar irradiation using artificial neural networks. Sol. Energy **82**, 181–187 (2008)
13. Jiang, Y.: Computation of monthly mean daily global solar. Energy **34**, 1276–1283 (2009)
14. Najafi, B., et al.: An intelligent artificial neural network-response surface methodology method. Energies **11**(4), 860 (2018)
15. Mathioulakis, E.: Artificial neural networks for the performance prediction of heat pump hot water heaters. Int. J. Sustain. Energ. **37**(2), 173–192 (2018)
16. Azeez, A.: Artificial neural network estimation of global solar. Appl. Sci. Res. **3**(2), 586–595 (2011)
17. Mosavi, A., et al.: Predicting the future using web knowledge: state of the art survey. In: Advances in Intelligent Systems and Computing, vol 660. Springer, Heidelberg (2018)
18. Chen, L.: Estimation of monthly solar radiation from measured temperatures using support vector machines-a case study. Renew. Energy **36**, 413–420 (2011)
19. Chen, J.L.: Assessing the potential of support vector machine for estimating daily solar radiation using sunshine duration. Energy Convers. Manag. **75**, 311–318 (2013)
20. Mosavi, A., Varkonyi-Koczy, A.R.: Integration of machine learning and optimization for robot learning. In: Advances in Intelligent Systems and Computing. Springer, Heidelberg (2017)

21. Chen, J.L., Li, G.S.: Evaluation of support vector machine for estimation of solar radiation from measured meteorological variables. Theor. Appl. Climatol. **115**, 627–638 (2014)
22. Guermoui, M.: Support vector regression methodology for estimating global solar radiation in Algeria. Eur. Phys. J. Plus **133**(1), 22 (2018)
23. Keshtegar, B.: Comparison of four heuristic regression techniques in solar radiation. Renew. Sustain. Energy Rev. **81**, 330–341 (2018)
24. Wu, J., Chan, C.K.: Prediction of hourly solar radiation using a novel hybrid model of ARMA and TDNN. Sol. Energy **85**, 808–817 (2011)
25. Moeini, I., et al.: Modeling the time-dependent characteristics of perovskite solar cells. Sol. Energy **170**, 969–973 (2018)
26. Mosavi, A., et al.: Industrial applications of big data: state of the art survey. Adv. Intell. Syst. Comput. **660**, 225–232 (2017)
27. Mosavi, A., et al.: Review on the usage of the multiobjective optimization package of modeFrontier in the energy. In: Advances in Intelligent Systems and Computing, pp. 217–224 (2017)
28. Mosavi, A., et al.: Reviewing the novel machine learning tools for materials design. In: Advances in Intelligent Systems and Computing, pp. 50–58 (2017)
29. Halabi, L.M.: Performance evaluation of hybrid adaptive neuro-fuzzy inference system models for predicting monthly global solar radiation. Appl. Energy **213**, 247–261 (2018)
30. Moeini, I., et al.: Modeling the detection efficiency in photodetectors with temperature-dependent mobility and carrier lifetime. In: Superlattices and Microstructures (2018)

A Parallel Fuzzy Filter Network for Pattern Recognition

Balazs Tusor[1]([⊠]), Annamária R. Várkonyi-Kóczy[2],
and József Bukor[3]

[1] Department of Mathematics and Informatics, J. Selye University,
Bratislavská Str. 3322, P.O. BOX 54, 945 01 Komárno, Slovakia
tusorb@ujs.sk
[2] Integrated Intelligent Systems Japanese-Hungarian Lab, Óbuda University,
Budapest, Hungary
[3] Institute of Automation, Óbuda University,
Bécsi Str. 96/b, Budapest 1034, Hungary

Abstract. Nowadays, parallelization is an increasingly popular tool to speed up algorithms. Data classification is one of the many fields of computer science that can take significant advantage of that. In this paper, a parallel implementation of Fuzzy RBF based filters are proposed for pattern recognition problems. It realizes a simple pattern matching by using the radial basis functions for proximity detection, then simply choosing the class or label associated to the pattern as output. The classifier has the advantage of being very simple to implement, to train and to modify the obtained knowledge. With the parallel computing improvement, the speed of both the training and evaluation phase are significantly increased compared to the sequential implementation.

Keywords: Classification · Data mining · Machine learning · Data structure
Pattern recognition · Parallel computing

1 Introduction

Classification is one of the most widely used fields of computer science. However, while nowadays the amount of training data is ever growing, so are the expectations towards the speed and performance of classifiers. Parallel computing [1] offers an effective solution to this problem, outsourcing the independent computational tasks into parallel processes, significantly accelerating the operation of such systems.

In this paper, a parallelized improvement of a classifier method is described. It is based on a modified fuzzy-RBF filter network that instead of calculating the activated linear combination of the input data, it realizes simple pattern matching by using the radial basis functions for proximity detection, then simply choosing the class or label associated to the pattern (characterized by the center and width parameters of the basis functions) as output. This classifier has the advantage of being very simple to implement, train (by clustering) and modify the trained knowledge (by adding or removing patterns), but it does not scale well with the amount and dimension of the input data.

G. Laukaitis (Ed.): INTER-ACADEMIA 2018, LNNS 53, pp. 275–282, 2019.
https://doi.org/10.1007/978-3-319-99834-3_36

With parallel computing, the speed of both the training and evaluation phase of the system can be significantly increased.

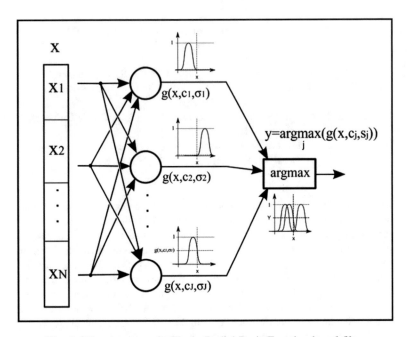

Fig. 1. The structure of a Fuzzy Radial Basis Function based filter

The rest of the paper is as follows. First, in Subsect. 2.1 the architecture of the base classifier is summarized. In Subsect. 2.2 the parallel improvements are described. The performance of the network is shown in Sect. 3. Finally, in Sect. 4 the conclusions are given.

2   Parallel F-RBF Based Filters

2.1   Fuzzy Radial Base Function Based Filter Networks

Radial Base Function networks [2] are neural networks where the neurons of the hidden layer evaluate a set of gaussian functions thus calculating proximity information between the input data and the center (c) parameter of each neuron. The output layer computes the weighted sum of the output of each neuron, thus producing the output of the system.

Fuzzy Radial Base Function based filter networks [4–8] are created from RBF networks by substituting the output layer with an arguments of the maxima (*argmax*) function. The idea is that each neuron represents a cluster in the problem space, and by this modification the task is to simply find the cluster that is the closest to the input data. Since the weights are no longer needed, a class label is assigned to each neuron instead (supervised learning is presumed, thus the class labels are available). Figure 1 illustrates the new structure, while (1) and (2) describe the applied formulas. The training of the system is done through clustering, where the cluster centers (c) and width (σ) parameters for each neuron are calculated. The former is gained directly from the input sample attributes (choosing a select few that can represent the rest), while the latter is calculated from the radius of each cluster, using (3) (which is derived from (2) by substituting 0.5 to the left side of the equation; meaning that it will $g()$ return values larger than 0.5 if the input sample is within the range of the given cluster).

$$y = argmax_j \left(g\left(x, c_j, \sigma_j\right)\right) \tag{1}$$

$$g\left(x, c_j, \sigma_j\right) = e^{-\frac{\|x-c_j\|^2}{2\sigma_j^2}} \tag{2}$$

$$\sigma_j = \frac{-r_j^2}{ln0.5} \tag{3}$$

2.2 Parallel Fuzzy RBF Based Filters

While the advantages of F-RBF filters are that they are simple to use and modify (as one only needs to add, remove or modify the parameters of the neurons), the main disadvantage is that they require a significant amount of computation, as all clusters are needed to be examined for each input sample. In case of image processing problems, this is a serious issue as typically there are at least hundreds of thousands of pixels in an image. One way to accelerate the training and operation is using parallel computing: the evaluation process can be applied to all pixels in the image at once, or at least as many pixels at once that are allowed by the parallel computing capabilities of the system.

In case of F-RBF based filters, it is presumed that there is a *background* or *unknown* class that the given patterns are needed to be distinguished from. In color filtering problems, this consists of the color tones that are marked as negative, i.e. not part of the color classes that are needed to be found in the image. While the background class is used during the training, it is unnecessary for the evaluation step of the classifier.

The training process of the classifier is illustrated in Fig. 2. The goal is to build a list of clusters with a manageable size that represents as much of the (non-background) training data as possible. The initial training list can potentially contain samples that are either redundant (having multiple instances of the same sample) or inconsistent (two samples from different classes having the same attributes) with each other, thus the first step is filtering these samples out. This can be done sequentially: comparing the samples one by one $(O(N^2)$ complexity); or in parallel. Depending on the size of the problem and the computational capabilities of the system, it can be fully parallel (if N^2 process can be launched at once, then it can be done with a technique called parallel reduction [9], O(*logN*)) or semi-parallel (*N* processes are launched, each compares a given sample to the others sequentially, O(*N*)).

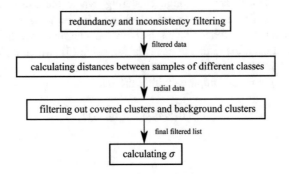

Fig. 2. The training process of parallel F-RBF based filters

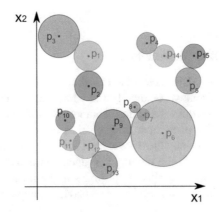

Fig. 3. A 2D example for multiclass clustering

The next step is creating circular clusters based on the filtered list: the cluster centers are derived from the attribute values, and the range of the clusters is set to half of the distance between the cluster center and the closest center that has a different class label. This is illustrated by Fig. 3, where each color corresponds to a given class. The radius of cluster *j* is given by

$$r_i = min(\{d(i,j)|k(i) \neq k(j)\}) \tag{4}$$

where $d(i, j)$ is the distance metric between clusters i and j, while $k(i)$ is the class label of cluster i.

Using the obtained radial data, the list of the clusters is filtered once more: the background class samples are filtered out, as well as clusters that are covered by larger clusters of the same class (e.g. in Fig. 3. p_7 is covered by p_6, thus it can be left out).

Finally, the σ parameter for each remaining cluster is calculated from the radial data (using (3)). This can be done in a single parallel step.

In the evaluation phase, the classifier can process the input data one by one in full parallel (one process for each cluster) or in batches (multiple inputs at once: one process per input). The latter is more typical for image processing problems like color pattern recognition, in which case the output (class label) for each pixel is calculated semi-parallel: one process for the combination of one pixel and all clusters, sequentially computing the values using (1) and (2). Furthermore, the resulting class label for each pixel is accepted only if the corresponding maximum value is larger than an arbitrary threshold parameter (Table 1).

Table 1. The time requirement, computation method and complexity of each step of the training. The computation method can be parallel (P), semi-parallel (SP) or sequential (S)

Operation	Required time	Computation	Complexity
Setting up the list of centers	1.06246 ms	P	$O(1)$
Flagging redundancy to reduce the data	13898 ms	SP	$O(N_1)$
Filtering out flagged data	329.7 ms	S	$O(N_1)$
Calculating radial information	672.832 ms	SP	$O(N_2)$
Filtering out background and covered clusters	161.84 ms	S	$O(N_2)$
Calculating σ parameters	0.0182 ms	P	$O(1)$

3 Experimental Results

The performance of the system is evaluated on an average desktop computer (Intel® Core™ i5-4590 CPU @ 3.30 GHz, 16 GB RAM, Gigabyte® GeForce™ GTX 960 graphic card with 4 GB GDDR5 RAM) using MS Visual Studio 2015, CUDA 9.0 and OpenCV 3.3.0.

In the experiment, the task is to recognize areas of images based on color information, e.g. areas with human skin tones. Figure 4 (left) shows a training image (with resolution 640×480) and (right) highlights the 3 distinguished classes (yellow: skin region, red: chair, magenta: bin). The rest of the image is marked as the background color; thus, the input data consists of $N_1 = 307200$ input data values. For color representation, RGB is used because the distance between two slightly different shades of the same color is typically much smaller than in the HSV color space.

The available configuration provides only a limited number of processes that can be launched simultaneously, thus some steps of the training and the evaluation step are implemented in a semi-parallel way.

First, the initial array is set up form the input image (parallel, ~ 1 ms). Then the redundant or inconsistent data are flagged (semi-parallel, ~ 13898 ms = 13.9 s). The data is then filtered (sequentially ~ 329 ms), and thus a reduced list is built with only $N_2 = 36998$ samples ($\sim 88\%$ reduction).

After that, the radiuses are calculated for each cluster (semi-parallel, ~ 672 ms), then background and covered clusters are marked and filtered out (sequentially,

Fig. 4. The training image (left) and mask (right) with the 3 classes marked

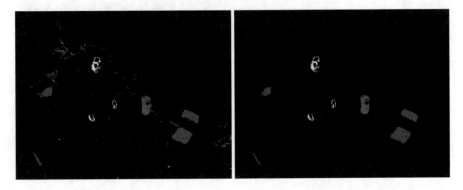

Fig. 5. The results of the training for thresholds 0.5 (left) and 0.9 (right)

~ 161.84 ms), resulting in the final $N_3 = 4118$ clusters. Finally, the σ parameters are calculated (parallel, ~ 0.0182 ms).

The classifier takes ~ 1.2 s to process one image (640×480, 4118 cluster comparisons per pixel). Figure 5 shows the results for thresholds 0.5 (left) and 0.9 (right). As it can be seen, the sensitivity of the classifier can be tuned with the threshold parameter. Some skin areas are also not found because their color is too similar to background color tones, thus they were removed during the training. The operational complexity is $O(N_3)$, which is a significant reduction from the complexity of the sequential version ($O(N_1 N_3)$).

The speed of the trained classifier was also tested on a single (non-batched) input, where the parallel operation can be fully exploited (each process computing a comparison to a single cluster). The time required to evaluate a single input (a color tone) takes ~ 0.025 ms.

4   Conclusions

In this paper, a parallel implementation of Fuzzy RBF based filters are proposed for pattern recognition problems. It realizes a simple pattern matching by using the radial basis functions for proximity detection, then simply choosing the class or label associated to the pattern as output. This classifier has the advantage of being very simple to implement, train by clustering and modify the trained knowledge. With the parallel computing improvement, the speed of both the training and evaluation phase are significantly increased compared to the sequential implementation, as the complexity is reduced from $O(N^2)$ to $O(N)$ during the training (with N training data), and from $O(PN)$ to $O(P)$ (considering P clusters) during the evaluation phase.

While the classifier is faster than the sequential version, there are numerous options to improve its speed. For example, the filtering of the lists can be done with a technique called parallel prefix sum [10], potentially lowering the complexity from $O(N)$ to $O(logN)$. The most time-consuming part of the training is the redundancy and inconsistency reduction step, which could be sped up with parallel reduction. In future work, these options will be further explored and implemented.

Acknowledgements. This work has partially been sponsored by the Research & Development Operational Program for the project "Modernization and Improvement of Technical Infrastructure for Research and Development of J. Selye University in the Fields of Nanotechnology and Intelligent Space", ITMS 26210120042, co-funded by the European Regional Development Fund.

References

1. Rumelhart, D.E., McClelland, J.: Parallel Distributed Processing: Explorations in the Microstructure of Cognition. MIT Press, Cambridge (1986)
2. Broomhead, D.S., Lowe, D.: Radial basis functions, multi-variable functional interpolation and adaptive networks. Technical report, RSRE. 4148 (1988)
3. Campbell-Kelly, M., Croarken, M., Robson, E.: The History of Mathematical Tables from Sumer to Spreadsheets, 1st ed. New York (2003)
4. Tusor, B., Várkonyi-Kóczy, A.R.: A hybrid fuzzy-RBFN filter for data classification. Adv. Mater. Res. **1117**, 261–264 (2015)
5. Várkonyi-Kóczy, A.R., Tusor, B., Bukor, J.: Data classification based on fuzzy-RBF networks. In: Proceedings of the 6th International Workshop on Soft Computing Applications, Timişoara, Romania, 24–26 July 2014 (2014)
6. Tusor, B., Várkonyi-Kóczy, A.R.: A hybrid fuzzy-RBFN filter for data classification. In: Proceedings of the 13th International Conference on Global Research and Education in Intelligent Systems, Interacademia 2014, Riga, Latvia, 9–12 September 2014 (2014)
7. Tusor, B., Várkonyi-Kóczy, A.R.: A rule-based filter network for multiclass data classification. In: Proceedings of the 2015 IEEE International Instrumentation and Measurement Technology Conference, I2MTC 2015, Pisa, Italy, 11–14 May 2015, pp. 1102–1107 (2015)

8. Várkonyi-Kóczy, A.R., Tusor, B., Bukor, J.: Data classification based on fuzzy-RBF networks. In: Balas, V.E., Jain, L.C., Kovačević, B. (eds.) Advances in Intelligent Systems and Computing, vol. 357, pp. 829–840. Springer, Berlin (2015)

9. Chandra, R.: Parallel Programming in OpenMP, pp. 59–77. Morgan Kaufmann, Los Altos (2001)

10. Cole, R., Vishkin, U.: Deterministic coin tossing with applications to optimal parallel list ranking. Inf. Control **70**(1), 32–53 (1986)

Numerical Simulation of Shaking Optimization in a Suspension Culture of iPS Cells

Kelum Elvitigala[1], Yoshiki Kanemaru[1], Masaki Yano[1],
Atsushi Sekimoto[1], Yasunori Okano[1(✉)], and Masahiro Kino-Oka[2]

[1] Department of Material Engineering Science, Osaka University,
1-3 Machikaneyama, Toyonaka, Osaka 560-8531, Japan
Okano@cheng.es.osaka-u.ac.jp
[2] Department of Biotechnology, Osaka University, 1-2 Yamadaoka, Suita,
Osaka 565-0971, Japan

Abstract. Research on induced Pluripotent Stem (iPS) cell has attracted attention due to their remarkable progress in regenerative medicine. Cost and quality are two main factors to be considered when using iPS cells in biological applications in a large scale. In this study, the suspension culture of iPS cells was numerically simulated in a cylindrical tank under the two-different shaking methods; one-direction rotation and periodic alternate rotation. The two shaking methods exhibited a significant difference in the average number of cells accumulated in the bottom. Even though the one direction rotation method suppressed the shear stress acting on iPS cells, after eight seconds it accumulated more cells in the bottom than the periodic alternate rotation method.

Keywords: iPS cell · Numerical simulation · Suspension culture

1 Introduction

At present, as a task for practical applications of iPS cells for the regenerative medicine, establishment of large-scale culture device is required. For the biomedical applications such as substitute human organs, 108–109 of iPS cells are demanded [1]. Therefore, practical applications of induced Pluripotent Stem (iPS) cells require the development of new automated culture method with keeping the undifferentiated state of iPS cells. To achieve this, it is necessary to automatically operate within one of the suspension cultures by replacing the manual static culture. Culturing in the laboratories depend on manual work by experienced technicians and it is not feasible to produce a large quantity of iPS cells. As a specific solution to solve this problem, it is necessary to change from the static culture in which cells are cultured two-dimensionally on the present culture dish to suspension in which cells are cultured three-dimensionally in a shaking tank. However, it has been reported that shear stress that is caused by shaking promotes the differentiation of mouse embryonic stem cells and it is expected that iPS cells also will similarly behave [2, 3]. An orbital shaking tank has been widely utilized in bioengineering and the shaking speed and the shaking methods have optimized [4]. Although it is difficult to measure the influence of above parameters on the cell in actual experiments, it can be easily predicted using computer simulation. In this study,

© Springer Nature Switzerland AG 2019
G. Laukaitis (Ed.): INTER-ACADEMIA 2018, LNNS 53, pp. 283–289, 2019.
https://doi.org/10.1007/978-3-319-99834-3_37

a numerical investigation was carried out for two different shaking methods, the one direction rotation (ODR) and the periodic alternate rotation (PAR), to obtain the average and maximum shear stress and sedimentation rate in a shaking tank.

2   Numerical Analysis and Numerical Procedure

In the simulation, cell colonies were modeled as rigid sphere particles and it was assumed that shape does not change during the simulation. The discrete element method (DEM) was used to study the dynamics of the particles and computational fluid dynamics (CFD) was applied to examine the dynamic of the culture fluid [4, 5]. In this numerical analysis, DEM was coupled with CFD to investigate the dynamics of the particles and the following assumptions were made in the simulation;

- The collision between particles is negligible as the number of particles is very small.
- The fluid is Newtonian and incompressible.
- Any biological reaction does not occur during the simulation.
- The temperature of the medium and physical properties involved remain constant during the process.
- The behavior of particles does not affect on the fluid flow.
- Particles Reynolds number is almost zero (10 rpm).

The numerical domain is shown in Fig. 1. The container was circumferentially shaken while keeping the horizontal configuration.

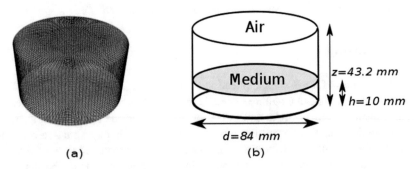

Fig. 1. Schematics of the (a) numerical grid used in the simulation and (b) numerical domain in an orbital shaking tank

In Fig. 1, d is diameter of the shaking tank, z is height of the shaking tank, and h is initial height of the liquid.

Figure 2 shows the shaking system and rotational directions of ODR and PAR. The rotation radius R_s of the both methods (ODR and PAR) was kept constant. Here ω is the angular velocity.

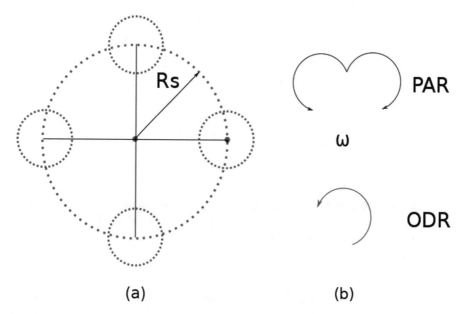

Fig. 2. Schematics of the (a) shaking system and (b) rotational directions of both ODR and PAR methods

Solid Phase (DEM equation). In this study, DEM is used to calculate the movement of particle and the motion of each particle is governed by the Newton's second law of motion.

$$m_s \frac{\partial v_s}{\partial t} = F_D + F_G - m_s \frac{\partial U}{\partial t} \tag{1}$$

$$F_D = C_D A_p \frac{\rho_f |V_f - V_s|^2}{2} \frac{V_f - V_s}{|V_f - V_s|} \tag{2}$$

$$C_D = \begin{cases} \frac{24}{Re}\left(1 + \frac{1}{6}Re^{\frac{2}{3}}\right) & Re \leq 1000 \\ 0.424 & Re \geq 1000 \end{cases} \tag{3}$$

$$Re = \frac{\rho_f |v_f - v_s| d_s}{\mu_f} \tag{4}$$

Here, subscript f and s represent the fluid and solid phases, respectively, and F_D is fluid resistance and F_G is buoyancy. m is mass, v is velocity, U is velocity of the shaking plate, t is time, μ is viscosity and ρ is density. C_D is the drag coefficient, A_p is the projected area. Re refers to the Reynolds number, d_s represent the diameter of the particle.

Liquid Phase. CFD-DEM were coupled with volume of fluid (VOF) method for the calculations in the fluid phase. The governing equations are the following continuity and Naiver Stokes equations;

$$\frac{\partial \rho_f}{\partial t} + \nabla\left(\rho_f v_s\right) = 0 \tag{5}$$

$$\frac{\partial v_f}{\partial t} v_f \cdot \nabla v_f = -\frac{1}{\rho_f} \nabla p + v_f \nabla^2 v_f + F_\sigma \tag{6}$$

$$\mathbf{F}_\sigma = \sigma k n \delta \tag{7}$$

F_σ stands for the surface tension term, σ is the surface tension, k is the interface curvature, n is the normal vector and δ is stands for delta function [6]. The velocity of the orbital shaking wall in both (ODR and PAR) methods are expressed as follows;

ODR Method;

$$U = \begin{pmatrix} U_x \\ U_y \\ U_z \end{pmatrix} = \begin{pmatrix} -R_s \omega \sin \omega t \\ -R_s \omega \cos \omega t \\ 0 \end{pmatrix} \tag{8}$$

PAR Method;

$$U = \begin{pmatrix} U_x \\ U_y \\ U_z \end{pmatrix} = \begin{pmatrix} \frac{\pi}{2} R_s \omega \sin\left(\pi \cos \frac{\pi}{2} t\right) \sin \frac{\omega}{2} t \\ \frac{\pi}{2} R_s \omega \cos\left(\pi \cos \frac{\pi}{2} t\right) \sin \frac{\omega}{2} t \\ 0 \end{pmatrix} \tag{9}$$

In addition, the following VOF method was used to capture the gas-liquid interface and it is governed by the following transport equation;

$$\frac{\partial \alpha}{\partial t} + \nabla \cdot (\alpha v) + \nabla \cdot ((1 - \alpha)\alpha \cdot v_r) = 0 \tag{10}$$

Fraction function α is defined as, $\alpha = 1$ and $\alpha = 0$ for liquid and air respectively. For the interface, value of the α can be expressed as $0 < \alpha < 1$.

$$v = \alpha v_t + (1 - \alpha)v_g \tag{11}$$

$$v_r = v_l - v_g \tag{12}$$

Here, subscript l and g represent the liquid and the gas respectively. In the computation, Eqs. (5)–(12) were discretized in the framework OpenFOAM using finite volume method (FVM). Velocity and pressure fields were coupled by the PISO algorithm [7]. The discretization scheme for the interface curvature was done by the second-order linear interpolation. The first order implicit Euler method was applied to the discretization for time derivatives.

3   Calculation Results and Discussion

3.1   Shear Stress

Shear stress acting on all the particles is discussed. Slip velocity (V_{slip}) acting on particle was used to evaluate the shear stress. The slip velocity was calculated by the velocity difference between the particle and the fluid as follows;

$$\left|V_{slip}\right| = \left|V_f - V_l\right| \tag{13}$$

As the dynamic viscosity and particle diameter are constant, the magnitude of the slip velocity indirectly indicates the magnitude of shear stress on the particle. We consider the average and maximum slip velocity acting on the particles. The average slip velocity is defined as;

$$\left|V_{slip}\right| = \frac{\left(\sum_{i=1}^{n}\left|V_{slip,i}\right|\right)}{n} \tag{14}$$

The magnitude of the maximum slip velocity $\left|V_{slip,max}\right|$ was defined as the maximum shear stress on all the particles. In the study, the angular velocity of the both models were set to 3.97 rad/s^{-1} and 3.24 rad/s^{-1} in ODR and PAR methods respectively to obtain the average slip velocity. As shown in the Figs. 3 and 4, it is evident that the average and maximum shear stress values were suppressed in the ODR method. In order to reduce the shear stress in PAR method, rotational velocity must be changed. Also, the diameter ratio Dr was defined by using the turning radius and the inner diameter d of the agitation tank as follows;

$$Dr = \frac{2R_s}{d} \tag{15}$$

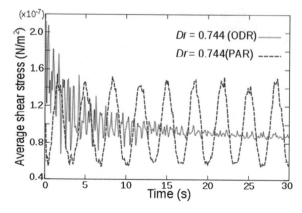

Fig. 3. Time development of average shear stress in both ODR and PAR methods

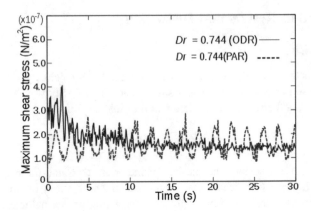

Fig. 4. Time development of maximum shear stress in both ODR and PAR methods

As shown in Figs. 3 and 4, it was found that both the average value and the maximum value of shear stress of PAR method oscillated in accordance with the cycle of rotation, and it can be confirmed that the amplitude is much larger than the ODR method. It can be expected that oscillation of shear stress has a high impedance to high-quality cell culture when iPS cell promotes differentiation and cell death even by instantaneous strong shear stress. It is thought that improvement to suppress these features and development of new shaking methods are necessary.

3.2   Sedimentation of the Particles

The problem associated with the particle sedimentation of the bottom is the prevention of oxygen transfer through the cell colonies. Also, the collision staying at the bottom of the tank leads to the cell differentiation. As shown in the Fig. 5, the sedimentation of the particles at the bottom is roughly the same in both methods until 8 s. After 8 s, the

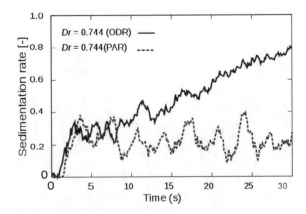

Fig. 5. Time development of particle sedimentation at the bottom of the tank in both ODR and PAR method

sedimentation increases in the ODR method. We can suggest that the high vertical velocity generated in PAR method compared with ODR method reduced the sedimentation of particles at the bottom.

4 Conclusion

We have developed a particle analysis model in a shaking tank that performs one direction rotation and periodic alternate rotation using OpenFOAM, and the shear stress applied to the particles and sedimentation of the particles were evaluated. It was found that the one direction rotation (ODR) method reduced the average and maximum shear stress compared with the periodic alternate rotation (PAR) method. However, it was found that ODR method cannot reduce the collision frequency of the cells in the bottom of tank. By introducing the PAR method, the collision frequency of the cells in the bottom of the tank can be reduced. Especially in this research, the following knowledge was obtained regarding periodic alternate rotation method; under a constant shear stress, the sedimentation ratio of the PAR method having the minimum value compared with the ODR method. Since the shear stress applied to the particles, greatly enhance according to the cycles of rotation, it is necessary to develop a new shear stress suppression method.

Acknowledgements. This research work was financially supported by the project of "Development of Cell Production and Processing System for Commercialization of Regenerative Medicine" from Japan Agency for Medical Research and Development, AMED, and by Grant-in-Aid for Scientific Research (B) (JSPS KAKENHI Grant Number JP15H04173) from Ministry of Education, Culture, Sports, Science and Technology of Japan, and supported partly by Collaborate Research Program for Young Scientists of ACCMS and IIMC, Kyoto University.

References

1. Olmer, R., et al.: Suspension culture of human pluripotent stem cells in controlled, stirred bioreactors. Tissue Eng. Part C Methods **18**(10), 772–784 (2012)
2. Adamo, L., et al.: Biomechanical forces promote embryonic haematopoiesis. Nature **459** (7250), 1131–1135 (2009)
3. Yamamoto, K., et al.: Fluid shear stress induces differentiation of Flk-1-positive embryonic stem cells into vascular endothelial cells in vitro. Am. J. Physiol. Circ. Physiol. **288**(4), H1915–H1924 (2005)
4. Yamamoto, T., Yano, M., Okano, Y., Kino-oka, M.: Numerical investigation for the movement of cell colonies in bioreactors: stirring and orbital shaking tanks. J. Chem. Eng. Japan **51**(5), 423–430 (2018)
5. O'Sullivan, C.: Particulate Discrete Element Modelling. CRC Press, Boca Raton (2014)
6. Brackbill, J., Kothe, D., Zemach, C.: A continuum method for modeling surface tension. J. Comput. Phys. **100**(2), 335–354 (1992)
7. Issa, R.: Solution of the implicitly discretised fluid flow equations by operator-splitting. J. Comput. Phys. **62**(1), 40–65 (1986)

The Flip-Side of Academic English (AE)

Valerie A. Wilkinson[1]([✉]), Damon M. Chandler[2],
and Takashi Mashiko[2]

[1] Faculty of Informatics,
Shizuoka University, Hamamatsu, Shizuoka 432-8561, Japan
vwilk@inf.shizuoka.ac.jp
[2] Faculty of Engineering, Shizuoka University,
Hamamatsu, Shizuoka 432-8561, Japan

Abstract. While we reported positive changes in class atmosphere in "Flipping Out in Japan" at iA2017 in Iasi, Romania, questions regarding empirical evidence of language improvement were lacking. Also, the control and experimental groups were not adequately demarcated. Our evaluations were subjective and impressionistic. With the addition of a third partner who advised a final testing protocol, we hope to address both of these criticisms with replicable test results. In this article, we re-examine our premises and the goals of our research. The "flipped classroom," or experiential learning, is all about context, and because both control and experimental groups did well last year in the class with a decidedly improved atmosphere, we decided this year to maintain the "flipped" format for both groups. With the enhancements of a senior student "mentor" and visitors to the experimental group, and two native English speakers teaching the control group, we are relying on a series of quizzes structured on the relevant material in each module of the class, with the final exam based on those quizzes, to provide robust empirical data which will convince peers of the approach's utility.

Keywords: Situated learning · Group dynamics · Project based learning
Cooperative principle

1 Introduction

While we reported positive changes in class atmosphere in "Flipping Out in Japan" at iA2017 in Iasi, Romania, we lacked empirical evidence to demonstrate language improvement. Also, we could not avoid contaminating the control and experimental groups. Our evaluations were subjective and impressionistic. With the addition of a third partner who advised a final testing protocol, we hope to address both of these criticisms with replicable test results. In the article, we re-examine our premises and the goals of our research. The "flipped classroom", or experiential learning, is all about context. Since both control and experimental groups did well in the class with a decidedly improved atmosphere, we decided to maintain the "flipped" format for both groups. With the enhancements of a senior student "mentor" and visitors to the experimental group and two "native speakers of English" teaching the control group, we are relying on a series of quizzes structured on the relevant material in each module

© Springer Nature Switzerland AG 2019
G. Laukaitis (Ed.): INTER-ACADEMIA 2018, LNNS 53, pp. 290–296, 2019.
https://doi.org/10.1007/978-3-319-99834-3_38

of the class, with the final exam based on those quizzes, to provide the robust empirical data which will convince our judges of the validity of the approach.

2 Background and Motivation

Japanese students are quite adept at self-regulated learning (SRL), particularly when it comes to topics in STEM. In fact, the Japanese university framework seems to have evolved to rely heavily on students' SRL skills. Classes meet only once a week for 90 min, where the material is delivered in a lecture-type setting to masses of students; the impetus is placed on the students' SRL skills to learn via homework. This symbiosis between teaching and SRL is effective for all subjects that a Japanese undergraduate will encounter—with one notable exception: language which, in our case, means English. English is required in STEM subjects.

When it comes to English as a foreign language (EFL), Japanese students have been showing progressively declining scores for years [4]. Researchers have investigated numerous possible causes—teacher efficacy, cultural differences, student motivation, anxiety, attitude, and willingness to communicate in EFL settings [1, 2, 9, 11–14] - with largely inconclusive findings.

We suspect a more fundamental cause: Meaningful EFL learning requires high-quality, high-quantity exposure [5], particularly with frequent speaking and listening opportunities. Here the Japanese university teaching framework breaks down at its foundation: The infrequency of classes, the large class sizes, and the inability to lean on SRL, makes developing EFL proficiency a largely unattainable goal.

Because Japanese students do not know how to study English speaking/listening, yet they are compelled to study something, they usually fall back to rote memorization of words and grammar. Indeed, Japanese people know many English words—including numerous English words that are used in Japanese culture—Japanese people are infamous for their inability to speak English and understand spoken English. Furthermore, without extensive editing, written English in Japan is largely incomprehensible. Clearly, knowing vocabulary and being able to use that vocabulary to communicate effectively are two very different challenges that require very different approaches. Actively forming one's own thoughts into intelligible sentences and then speaking those sentences is a crucial experience for language acquisition, but is one that has been ignored in Japan.

Another problem is the fact that language learning is naturally slow; it is extremely difficult to notice one's own progress and therefore equally difficult to maintain the persistence required for true English proficiency. This perceived lack of progress, with no apparent solution, ultimately leads to severe demotivation and/or abandonment [10]. Unless radical changes in English pedagogy are made, Japan will continue as a country in which English proficiency remains a fanciful goal.

3 Our Proposed Study

We stand at a unique intersection of STEM and EFL. Each year, we teach Academic English to over 100 third-year Japanese engineering students at Shizuoka University. The goal of the course is for the students to be able communicate with foreign engineers with enough proficiency to solve engineering challenges. This goal is similar to normal EFL goals, but the engineering subject matter holds special value on two fronts: (1) The students are naturally inclined to the topics; and (2) English and Japanese engineering languages share a rooting in the underlying STEM concepts.

By using this course as a testbed, we propose to research the effects of "the flipped classroom" or experiential learning—using a platform that sparks an inner drive to innovate—on student ability, mindset, motivation, and willingness to use English. The very few studies on Project Based Learning (PBL) for EFL in Japan have yielded inconclusive results [3, 6], with the limited benefits attributable to irrelevant projects and the lack of out-of-class contact with English speakers. Those studies did not take account of the kind of natural group dynamic processes which have been studied and analyzed extensively since Tuckman's [15] "Development Sequence in small groups," the resonance possible in communities of practice (COP), since Lave and Wenger proposed the COP concept [7], and Grice's Cooperative Principle [4], from the field of linguistics and pragmatics, suggesting that people have a tacit understanding of how discussions ought to proceed. We trace these matters as a blueprint, suggesting how team development works. Energetics and group dynamics are not, strictly speaking, part of the class, but class environment promotes a better atmosphere for development.

4 The Project

Mini-Conference Poster sessions are centered around a Robotics theme. This theme will be effective for Electronic, Mechanical, Systems, and Materials Engineers. The interdisciplinary strength of Robotics ensures that most engineering students will be able to connect directly with the topic. The focus of the major provides the focus of the themes. Electrical Engineers use boards, circuits, motors, programs, sensors, and batteries. Through cooperative work in teams, each group will be able to achieve commonality of purpose and coordination of their efforts.

5 Details of the Research Plan and Methods

The course on which this study is based is Academic English for Engineering, aimed at third-year Electrical and Electronic Engineering undergraduates. We have been using a flipped course design with in-class exercises, out-of-class Moodle exercises, and small projects. For the proposed study, the mini-poster themes will be structured around aspects of building a robot.

5.1   Set up "Control Group" and "Experimental Group"

Control Group. The class is divided into groups of three or four. The room set-up precludes groups of five. The teachers are two Americans, the Engineering professor and the language facilitator (PI).

Experimental Group. The class is divided into teams and instructed to select a team leader or representative. The "coach" assists with leader conferences, translates explanation presentation into Japanese, cruises the class for questions, confers with professor and interprets student queries. The poster presentations in the second and third week of each session are open to visitors to participate as an audience with questions.

5.2   The Sessions

Session One. Select Robot Contest to Enter. 1st week: Class given basic instructions to do search and parameters for discovery. Homework: prepare slides. 2nd and 3rd weeks: "Mini-poster conference". Homework: take session quiz.

Session Two. Power, Motors, Body. Homework: prepare slides. 2nd and 3rd weeks: "Mini-poster conference". Homework: take session quiz.

Session Three. Software and Communications. Homework: prepare slides. 2nd and 3rd weeks: "Mini-poster conference". Homework: take session quiz.

Session Four. Machine Intelligence: Bridge to Research. Homework: prepare slides. 2nd and 3rd weeks: "Mini-poster conference". Homework: take session quiz.

5.3   Use of Moodle Platform

The Moodle provides each student with a use ID. It is used to take attendance, post the Session Theme PowerPoint files, and administer quizzes.

5.4   Testing

Records of quizzes and final exams produce the comparative information to assess whether the structure of the control and experimental groups contributed to a significant difference or not.

Session Quizzes. A short composition based on each session theme is used to give a comprehension test. These quizzes are posted on the Moodle.

Final Exam. The final exam is based on the four session quizzes. Access to all four compositions gives students an opportunity to prepare.

6 Discussion

The structure of the experiment divides the class into control and experimental groups. This is based on whether the students are electrical or electronic majors. The room available has certain inflexible features.

6.1 Project Based Learning (PBL)

The mini-poster conference structure is perfect for PBL. The four iterations give students a usable pattern for dividing the project into jobs, doing the preparation at home, and being responsible for their poster in the presentation session.

6.2 "The Knowledge Creating Classroom"

The *Knowledge Creating Company* [8] posits that middle-up-down management is productive of innovation and creativity, as compared to the traditional top-down or bottom-up structures. "Middle," in the case of companies, refers to middle management. We propose that the university is also a knowledge creating institution. The projects of students preparing for graduation and the work of graduate students in the lab are "knowledge creation" itself. The selection of the student coach is important. The coach should be invested in the process, "like a TA", who has access to students and professors. He (or she) is developing self-image and mentoring abilities while providing an attractive role model for the younger students. The "workshop" feel of the classroom conveys intrinsic information to students. The professor is available to discuss points in detail. The mood is optimistic and awake.

6.3 Group Dynamics, Situated Learning, and Cooperative Principle

Many years of language teaching and exposure to multiple methods convinced the PI that active and experiential learning that engages the whole person are the best. In 1965, Tuckman first proposed the "Developmental sequence in small groups" [15] which became the well-known "5 phases of group life". "Forming, Storming, Norming, Performing, Adjourning". The process is a description of group life which plays out in real contexts. Some questions are: What is the best technique of group selection? What strategies can be employed for smooth performance?

All the students are working together in their teams. Hopefully all are purposefully drawn into the work of the group. They learn from each other and learn by doing. In such a situation, the thesis statement contained in the subtitle of Lave and Wenger's [7] *Situated Learning* makes good sense. "Legitimate peripheral participation" is an organismic, sustainable way of acclimatizing to the work-focused community of practice.

Finally, Grice's work with the cooperative principle describes practices that discussion groups appear to obey, without making them explicit. This paper is not about those descriptions of group dynamics, process, and progress. The class is not "about" those concepts, either. If we structure the class with those principles in mind, it might make a productive difference to the team, group, and class.

7   Conclusions

In this article, we have described a study (an extension of our research presented at last year's IA meeting) designed to investigate the effects of a "flipped classroom" using mini poster sessions for university-level Japanese learners of academic English. The class was divided into control and test groups, in which both groups were exposed to two native English-speaking instructors, but in which only the test group had access to a senior student mentor and visitors who could create a much more engaging atmosphere. Via a series of theme-based quizzes (currently being administered), and a final exam based on those quizzes, we will discuss at the IA meeting the differences between the groups. In the future, we hope to perform a follow-up study in which the course is taught to other students (e.g., other technical majors), particularly students in countries other than Japan for comparative analysis.

References

1. Caprio, M.: University students and second language learning: a look at purpose and attitude. Acad. Lit. Lang. **42**, 23–36 (1987)
2. Dörnyei, Z.: Attitudes, orientations, and motivations in language learning: advances in theory, research, and applications. Lang. Learn. **53**(S1), 3–32 (2003)
3. Eguchi, M., Eguchi, K.: The limited effect of PBL on EFL learners: a case study of English magazine projects. Asian EFL J. **8**(3), 207–225 (2006)
4. EF Education First: EF English Proficiency Index (for Japan). https://www.ef.edu/epi/regions/asia/japan/. Accessed 20 June 2018
5. Hulstijn, J.H.: Intentional and incidental second language vocabulary learning: a reappraisal of elaboration, rehearsal and automaticity. In: Robinson, P. (ed.) Cognition and Second Language Instruction, pp. 258–286. Cambridge University Press, Cambridge (2001)
6. Kobayashi, M.: Second language socialization through an oral project presentation: Japanese university students' experience. In: Beckett, G.H., Chamness Miller, P. (eds.) Project-Based Second and Foreign Language Education: Past, Present and Future, pp. 71–93. Information Age, Greenwich (2006)
7. Lave, J., Wenger, E.: Situated Learning: Legitimate Peripheral Participation, pp. 89–113. Cambridge UP, Cambridge (1991)
8. Nonaka, I., Takeuchi, H.: The Knowledge-Creating Company. Oxford University Press, New York (1995)
9. Ryan, S.: Self and identity in L2 motivation in Japan: the ideal L2 self and Japanese learners of English. In: Dörnyei, Z., Ushioda, E. (eds.) Motivation, Language Identity and the L2 Self, pp. 120–143. Multilingual Matters, Clevedon (2009)
10. Sakai, H., Kikuchi, K.: An analysis of demotivators in the EFL classroom. System **37**(1), 57–69 (2009)
11. Sugita, M., Takeuchi, O.: What can teachers do to motivate their students? A classroom research on motivational strategy use in the Japanese EFL context. Innov. Lang. Learn. Teach. **4**, 21–35 (2010)
12. Tajino, A., Tajino, Y.: Native and non-native: what can they offer? Lessons from team-teaching in Japan. ELT J. **54**(1), 3–11 (2000)

13. Watanabe, Y.: External variables affecting language learning strategies of Japanese EFL learners: effects of entrance examination, years spent at college/university, and staying overseas. Master's thesis, Lancaster University, Lancaster, UK (1990)
14. Yashima, T.: Willingness to communicate in a Second Language: the Japanese EFL Context. Mod. Lang. J. **86**(i), 54–66 (2002)
15. Tuckman, B.W.: Developmental sequence in small groups. Psychol. Bull. **63**, 384–399 (1965)

Author Index

Printed in the United States
By Bookmasters